Relational
Data Clustering
Models, Algorithms,
and Applications

Chapman & Hall/CRC
Data Mining and Knowledge Discovery Series

SERIES EDITOR
Vipin Kumar
University of Minnesota
Department of Computer Science and Engineering
Minneapolis, Minnesota, U.S.A

AIMS AND SCOPE

This series aims to capture new developments and applications in data mining and knowledge discovery, while summarizing the computational tools and techniques useful in data analysis. This series encourages the integration of mathematical, statistical, and computational methods and techniques through the publication of a broad range of textbooks, reference works, and handbooks. The inclusion of concrete examples and applications is highly encouraged. The scope of the series includes, but is not limited to, titles in the areas of data mining and knowledge discovery methods and applications, modeling, algorithms, theory and foundations, data and knowledge visualization, data mining systems and tools, and privacy and security issues.

PUBLISHED TITLES

UNDERSTANDING COMPLEX DATASETS: DATA MINING WITH MATRIX DECOMPOSITIONS
David Skillicorn

COMPUTATIONAL METHODS OF FEATURE SELECTION
Huan Liu and Hiroshi Motoda

CONSTRAINED CLUSTERING: ADVANCES IN ALGORITHMS, THEORY, AND APPLICATIONS
Sugato Basu, Ian Davidson, and Kiri L. Wagstaff

KNOWLEDGE DISCOVERY FOR COUNTERTERRORISM AND LAW ENFORCEMENT
David Skillicorn

MULTIMEDIA DATA MINING: A SYSTEMATIC INTRODUCTION TO CONCEPTS AND THEORY
Zhongfei Zhang and Ruofei Zhang

NEXT GENERATION OF DATA MINING
Hillol Kargupta, Jiawei Han, Philip S. Yu, Rajeev Motwani, and Vipin Kumar

DATA MINING FOR DESIGN AND MARKETING
Yukio Ohsawa and Katsutoshi Yada

THE TOP TEN ALGORITHMS IN DATA MINING
Xindong Wu and Vipin Kumar

GEOGRAPHIC DATA MINING AND KNOWLEDGE DISCOVERY, SECOND EDITION
Harvey J. Miller and Jiawei Han

TEXT MINING: CLASSIFICATION, CLUSTERING, AND APPLICATIONS
Ashok N. Srivastava and Mehran Sahami

BIOLOGICAL DATA MINING
Jake Y. Chen and Stefano Lonardi

INFORMATION DISCOVERY ON ELECTRONIC HEALTH RECORDS
Vagelis Hristidis

TEMPORAL DATA MINING
Theophano Mitsa

RELATIONAL DATA CLUSTERING: MODELS, ALGORITHMS, AND APPLICATIONS
Bo Long, Zhongfei Zhang, and Philip S. Yu

Chapman & Hall/CRC
Data Mining and Knowledge Discovery Series

Relational Data Clustering

Models, Algorithms, and Applications

Bo Long
Zhongfei Zhang
Philip S. Yu

CRC Press
Taylor & Francis Group
Boca Raton London New York

CRC Press is an imprint of the
Taylor & Francis Group, an **informa** business
A CHAPMAN & HALL BOOK

CRC Press
Taylor & Francis Group
6000 Broken Sound Parkway NW, Suite 300
Boca Raton, FL 33487-2742

First issued in paperback 2019

© 2010 by Taylor & Francis Group, LLC
CRC Press is an imprint of Taylor & Francis Group, an Informa business

No claim to original U.S. Government works

ISBN-13: 978-1-4200-7261-7 (hbk)
ISBN-13: 978-0-367-38405-0 (pbk)

Library of Congress Cataloging-in-Publication Data

Relational data clustering : models, algorithms, and applications / Bo Long, Zhongfei Zhang, Philip S. Yu.
    p. cm. -- (Chapman & Hall/CRC data mining and knowledge discovery series)
    Includes bibliographical references and index.
    ISBN 978-1-4200-7261-7 (hardcover : alk. paper)
    1. Data mining. 2. Cluster analysis. 3. Relational databases. I. Long, Bo. II. Zhang, Zhongfei. III. Yu, Philip S. IV. Title. V. Series.

QA76.9.D343R46 2010
005.75'6--dc22   2010009487

Visit the Taylor & Francis Web site at
http://www.taylorandfrancis.com

and the CRC Press Web site at
http://www.crcpress.com

To my parents, Jingyu Li and Yinghua Long; my sister, Li; my wife Jing;
and my daughter, Helen
Bo Long

To my parents, Yukun Zhang and Ming Song; my sister, Xuefei; and my
sons, Henry and Andrew
Zhongfei (Mark) Zhang

To my family
Philip S. Yu

Contents

List of Tables   xi

List of Figures   xiii

Preface   xv

1 Introduction   1
   1.1   Defining the Area   1
   1.2   The Content and the Organization of This Book   4
   1.3   The Audience of This Book   6
   1.4   Further Readings   6

I   Models   9

2 Co-Clustering   11
   2.1   Introduction   11
   2.2   Related Work   12
   2.3   Model Formulation and Analysis   13
        2.3.1   Block Value Decomposition   13
        2.3.2   NBVD Method   17

3 Heterogeneous Relational Data Clustering   21
   3.1   Introduction   21
   3.2   Related Work   23
   3.3   Relation Summary Network Model   24

4 Homogeneous Relational Data Clustering   29
   4.1   Introduction   29
   4.2   Related Work   32
   4.3   Community Learning by Graph Approximation   33

5 General Relational Data Clustering   39
   5.1   Introduction   39
   5.2   Related Work   40
   5.3   Mixed Membership Relational Clustering   42
   5.4   Spectral Relational Clustering   45

6   Multiple-View Relational Data Clustering   **47**
  6.1   Introduction   47
  6.2   Related Work   49
  6.3   Background and Model Formulation   50
      6.3.1   A General Model for Multiple-View Unsupervised
            Learning   51
      6.3.2   Two Specific Models: Multiple-View Clustering and
            Multiple-View Spectral Embedding   53

7   Evolutionary Data Clustering   **57**
  7.1   Introduction   57
  7.2   Related Work   59
  7.3   Dirichlet Process Mixture Chain (DPChain)   60
      7.3.1   DPChain Representation   61
  7.4   HDP Evolutionary Clustering Model (HDP-EVO)   63
      7.4.1   HDP-EVO Representation   63
      7.4.2   Two-Level CRP for HDP-EVO   65
  7.5   Infinite Hierarchical Hidden Markov State Model   66
      7.5.1 iH^2MS Representation   67
      7.5.2   Extention of iH^2MS   68
      7.5.3   Maximum Likelihood Estimation of HTM   69
  7.6   HDP Incorporated with HTM (HDP-HTM)   70
      7.6.1   Model Representation   70

II   Algorithms   **73**

8   Co-Clustering   **75**
  8.1   Nonnegative Block Value Decomposition Algorithm   75
  8.2   Proof of the Correctness of the NBVD Algorithm   78

9   Heterogeneous Relational Data Clustering   **83**
  9.1   Relation Summary Network Algorithm   83
  9.2   A Unified View to Clustering Approaches   90
      9.2.1   Bipartite Spectral Graph Partitioning   90
      9.2.2   Binary Data Clustering with Feature Reduction   90
      9.2.3   Information-Theoretic Co-Clustering   91
      9.2.4   K-Means Clustering   92

10 Homogeneous Relational Data Clustering   **95**
  10.1 Hard CLGA Algorithm   95
  10.2 Soft CLGA Algorithm   97
  10.3 Balanced CLGA Algorithm   101

11 General Relational Data Clustering **105**

11.1 Mixed Membership Relational Clustering Algorithm 105

    11.1.1 MMRC with Exponential Families 105

    11.1.2 Monte Carlo E-Step 108

    11.1.3 M-Step 109

    11.1.4 Hard MMRC Algorithm 112

11.2 Spectral Relational Clustering Algorithm 114

11.3 A Unified View to Clustering 118

    11.3.1 Semi-Supervised Clustering 118

    11.3.2 Co-Clustering 119

    11.3.3 Graph Clustering 120

12 Multiple-View Relational Data Clustering **123**

12.1 Algorithm Derivation 123

    12.1.1 Multiple-View Clustering Algorithm 124

    12.1.2 Multiple-View Spectral Embedding Algorithm 127

12.2 Extensions and Discussions 129

    12.2.1 Evolutionary Clustering 129

    12.2.2 Unsupervised Learning with Side Information 130

13 Evolutionary Data Clustering **133**

13.1 DPChain Inference 133

13.2 HDP-EVO Inference 134

13.3 HDP-HTM Inference 136

III   Applications **139**

14 Co-Clustering **141**

14.1 Data Sets and Implementation Details 141

14.2 Evaluation Metricees 142

14.3 Results and Discussion 143

15 Heterogeneous Relational Data Clustering **147**

15.1 Data Sets and Parameter Setting 147

15.2 Results and Discussion 150

16 Homogeneous Relational Data Clustering **153**

16.1 Data Sets and Parameter Setting 153

16.2 Results and Discussion 155

17 General Relational Data Clustering **159**

17.1 Graph Clustering 159

17.2 Bi-clustering and Tri-Clustering 161

17.3 A Case Study on Actor-Movie Data 163

17.4 Spectral Relational Clustering Applications 164

    17.4.1 Clustering on Bi-Type Relational Data 164

  17.4.2 Clustering on Tri-Type Relational Data 166

18 Multiple-View and Evolutionary Data Clustering   169
  18.1 Multiple-View Clustering 169
      18.1.1 Synthetic Data 169
      18.1.2 Real Data 172
  18.2 Multiple-View Spectral Embedding 173
  18.3 Semi-Supervised Clustering 174
  18.4 Evolutionary Clustering 175

IV   Summary   179

References   185

Index   195

List of Tables

4.1   A list of variations of the CLGA model   37

9.1   A list of Bregman divergences and the corresponding convex functions   84

14.1 Data sets details. Each data set is randomly and evenly sampled from specific newsgroups   142
14.2 Both NBVD and NMF accurately recover the original clusters in the *CLASSIC3* data set   144
14.3 A normalized block value matrix on the *CLASSIS3* data set .   145
14.4 NBVD extracts the block structure more accurately than NMF on *Multi5* data set   145
14.5 NBVD shows clear improvements on the micro-averaged-precision values on different newsgroup data sets over other algorithms   146

15.1 Parameters and distributions for synthetic bipartite graphs .   148
15.2 Subsets of newsgroup data for constructing bipartite graphs .   148
15.3 Parameters and distributions for synthetic tripartite graphs .   149
15.4 Taxonomy structures of two data sets for constructing tripartite graphs   149
15.5 NMI scores of the algorithms on bipartite graphs   151
15.6 NMI scores of the algorithms on tripartite graphs   152

16.1 Summary of the graphs with general clusters   154
16.2 Summary of graphs based on text datasets   154
16.3 NMI scores on graphs of general clusters   155
16.4 NMI scores on graphs of text data   156

17.1 Summary of relational data for Graph Clustering   160
17.2 Subsets of newsgroup data for bi-type relational data   161
17.3 Taxonomy structures of two data sets for constructing tripartite relational data   161
17.4 Two clusters from actor-movie data   163
17.5 NMI comparisons of SRC, NC, and BSGP algorithms   166
17.6 Taxonomy structures for three data sets   167
17.7 NMI comparisons of SRC, MRK, and CBGC algorithms . . .   168

List of Tables

18.1 Distributions and parameters to generate syn2 data 170

List of Figures

1.1 Relationships among the different areas of relational data clustering. 3

2.1 The original data matrix (b) with a 2×2 block structure which is demonstrated by the permuted data matrix (a). The row-coefficient matrix R, the block value matrix B, and the column-coefficient matrix C give a reconstructed matrix (c) to approximate the original data matrix (b). 15
2.2 Illustration of the difference between BVD and SVD. 16

3.1 A bipartite graph (a) and its relation summary network (b). . 24
3.2 A tripartite graph (a) and its RSN (b) 26
3.3 The cluster structures of V_2 and V_3 affect the similarity between v_{11} and v_{12} through the hidden nodes. 28

4.1 A graph with mixed cluster structures (a) and its cluster prototype graph (b). 30
4.2 A graph with virtual nodes. 31
4.3 A graph with strongly connected clusters (a) and its cluster prototype graph (b); the graph affinity matrices for (a) and (b), (c) and (d), respectively. 34

5.1 Examples of the structures of relational data. 42

7.1 The DPChain model. 61
7.2 The HDP-EVO model. 64
7.3 The iH^{2}MS model. 67
7.4 The HDP-HTM model. 71

9.1 An RSN equivalent to k-means. 92

13.1 The illustrated example of global and local cluster correspondence 135

14.1 The coefficient of the variance for the columns of the mean block value matrix with the varing number of the word clusters using NBVD on different *NG20* data sets. 143

14.2 Micro-averaged-precision with the varing number of the word clusters using NBVD on different *NG20* data sets. 144

17.1 NMI comparison of SGP, METIS, and MMRC algorithms. . . . 160
17.2 NMI comparison among BSGP, RSN, and MMRC algorithms for bi-type data. 162
17.3 NMI comparison of CBGC, RSN, and MMRC algorithms for tri-type data. 162
17.4 (a), (b), and (c) are document embeddings of multi2 data set produced by NC, BSGP, and SRC, respectively (u_1 and u_2 denote first and second eigenvectors, respectively). (d) is an iteration curve for SRC. 165
17.5 Three pairs of embeddings of documents and categories for the TM1 data set produced by SRC with different weights: (a) and (b) with $w_a^{(12)} = 1, w_a^{(23)} = 1$; (c) and (d) with $w_a^{(12)} = 1, w_a^{(23)} = 0$; (e) and (f) with $w_a^{(12)} = 0, w_a^{(23)} = 1$. . . . 167

18.1 A toy example that demonstrates that our MVC algorithm is able to learn the consensus pattern from multiple-views with noise. 170
18.2 NMI comparison on synthetic data. 171
18.3 NMI comparison on real data. 172
18.4 Four embeddings for NGv3 data set. 174
18.5 NMI comparison on spectral embedding of NGv3 and NGv4. 175
18.6 Semi-supervised clustering results. 176
18.7 Evolutionary clustering results. 176

Preface

The world we live today is full of data with relations—the Internet, the social network, the telecommunications, the customer shopping patterns, as well as the micro-array data in bioinformatics research to just name a few examples, resulting in an active research area called relational data mining in the data mining research field. Given the fact that in many real-world applications we do not have the luxury to have any training data or it would become extremely expensive to obtain training data, relational data clustering has recently caught substantial attention from the related research communities and has thus emerged as a new and hot research topic in the area of relational data mining. This book is the very first monograph on the topic of relational data clustering written in a self-contained format. This book addresses both the fundamentals and the applications of relational data clustering, including the theoretic models, algorithms, as well as the exemplar applications of applying these models and algorithms to solve for real-world problems.

The authors of this book have been actively working on the topic of relational data clustering for years, and this book is the final culmination of their years of long research on this topic. This book may be used as a collection of research notes for researchers interested in the research on this topic, a reference book for practitioners or engineers, as well as a textbook for a graduate advanced seminar on the topic of relational data clustering. This book may also be used for an introductory course for graduate students or advanced undergraduate seniors. The references collected in this book may be used as further reading lists or references for the readers.

Due to the extensive attention received on this topic in the literature, and also due to the rapid development in the literature on this topic in recent years, it is by no means meant to be exhaustive to collect complete information on relational data clustering. We intend to collect the most recent research of our own on this topic in this book. For those who have already been in the area of relational data mining or who already know what this area is about, this book serves the purpose of a formal and systematic collection of part of the most recent advances of the research on this topic. For those who are beginners to the area of relational data mining, this book serves the purpose of a formal and systematic introduction to relational data clustering.

It is not possible for us to accomplish this book without the great support from a large group of people and organizations. In particular, we would like to thank the publisher—Taylor & Francis/CRC Press for giving us the opportunity to complete this book for the readers as one of the books in the Chapman

& Hall/CRC *Data Mining and Knowledge Discovery* series, with Prof Vipin Kumar at the University of Minnesota serving as the series editor. We would like to thank this book's editor of Taylor & Francis Group, Randi Cohen, for her enthusiastic and patient support, effort, and advice; the project coordinator of Taylor & Francis Group, Amber Donley, and the anonymous proof reader for their meticulous effort in correcting typos and other errors of the draft of the book; and Shashi Kumar of Glyph International for his prompt technical support in formatting the book. We would like to thank Prof Jiawei Han at University of Illinois at Urbana-Champaign and Prof Jieping Ye at Arizona State University as well as another anonymous reviewer for their painstaking effort to review the book and their valuable comments to substantially improve the quality of this book. While this book is derived from the original contributions by the authors of the book, part of the materials of this book are also jointly contributed by their colleagues Xiaoyun Wu at Google Research Labs and Tianbing Xu at SUNY Binghamton. This book project is supported in part by the National Science Foundation under grant IIS-0812114, managed by the program manager, Dr. Maria Zemankova. Any opinions, findings, and conclusions or recommendations expressed in this material are those of the authors and do not necessarily reflect the views of the National Science Foundation.

Finally, we would like to thank our families for the love and support that are essential for us to complete this book.

Chapter 1

Introduction

1.1 Defining the Area

Clustering problem is an essential problem of data mining and machine learning. Cluster analysis is a process that partitions a set of data objects into clusters in such a way that objects from the same cluster are similar and objects from different clusters are dissimilar [105].

Most clustering approaches in the literature focus on "flat" data, in which each data object is represented as a fixed-length attribute vector [105]. However, many real-world data sets are much richer in structure, involving objects of multiple types that are related to each other, such as documents and words in a text corpus; Web pages, search queries, and Web users in a Web search system; and shops, customers, suppliers, shareholders, and advertisement media in a marketing system. We refer those data, in which data objects are related to each other, as relational data.

Relational data have attracted more and more attention due to its phenomenal impact in various important applications, such as text analysis, recommendation systems, Web mining, online advertising, bioinformatics, citation analysis, and epidemiology. Different relational learning problems have been addressed in different fields of data mining. One of the most important relational learning tasks is to discover hidden groups (clusters) from relational data, i.e., relational data clustering. The following are examples of relational data clustering:

- Text analysis. To learn the document clusters and word clusters from the bi-type relational data, document-word data.

- Recommendation system. Movie recommendation based on user clusters (communities) and movie clusters learned from relational data involving users, movies, and actors/actresses.

- Online advertisement. Based on the relational data, in which advertiser, bidded terms, and words are interrelated to each other, the clusters of advertisers and bidder terms can be learned for bidded term suggestion.

- BBioinformatics. Automatically identifying gene groups (clusters) from the relational data of genes, conditions, and annotation words.

- Research community mining and topic identification. To identify research communities (author clusters) and research topics (paper clusters) from relational data consisting of authors, papers, and key words.

In general, relational data contain three types of information: attributes for individual objects, homogeneous relations between objects of the same type, and heterogeneous relations between objects of different types. For example, for a scientific publication relational data set of papers and authors, the personal information such as affiliations for authors is attributes; the citation relations among papers are homogeneous relations; the authorship relations between papers and authors are heterogeneous relations. Such data violate the classic IID assumption in machine learning and statistics and present huge challenges to traditional clustering approaches.

There are two frameworks for the challenging problem of relational data clustering. One is individual clustering framework; the other is collective clustering framework.

Under the individual clustering framework, we transform relational data into flat data and then cluster each type of objects individually. An intuitive way under this framework is to transform all the relations into features and then to apply traditional clustering algorithms directly. On the other hand, under the collective clustering framework, we cluster different types of data objects simultaneously. Compared with the collective clustering framework, individual clustering framework has the following disadvantages.

First, the transformation causes the loss of relation and structure information [48]. Second, traditional clustering approaches are unable to tackle influence propagation in clustering relational data, i.e., the hidden patterns of different types of objects could affect each other both directly and indirectly (pass along relation chains). Third, in some data mining applications, users are interested not only in the hidden structure for each type of objects, but also in interaction patterns involving multi-types of objects. For example, in document clustering, in addition to document clusters and word clusters, the relationship between document clusters and word clusters is also useful information. It is difficult to discover such interaction patterns by clustering each type of objects individually.

One the other hand, the collective clustering framework has the obvious advantage of learning both local and global cluster structures based on all three types of information. In this book, our main focus is on the collective clustering framework. Under the collective clustering framework, with different foci on different types of relational data, there are different subfields of relational data clustering: co-clustering on bi-type heterogeneous relational data, heterogeneous relational data clustering on multi-type heterogeneous relational data, homogeneous relational data clustering on homogeneous relational data, general relational data clustering on general relational data.

Another interesting observation is that a number of important clustering problems, which have been of intensive interest in the literature, can be viewed

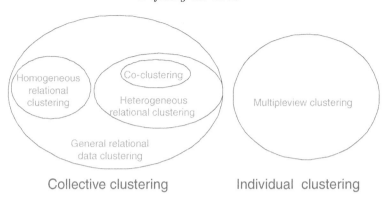

FIGURE 1.1: Relationships among the different areas of relational data clustering.

as special cases of relational clustering. For example, graph clustering partitioning [28, 75, 113] can be viewed as clustering on singly ype relational data consisting of only homogeneous relations (represented as a graph affinity matrix) co-clustering [11, 44], which arises in important applications such as document clustering and micro-array data clustering; under the collective clustering framework, this can be formulated as clustering on bi-type relational data consisting of only heterogeneous relations. Recently, semi-supervised clustering [14, 124] has attracted significant attention, which is a special type of clustering using both labeled and unlabeled data. In Section 11.3, we show that semi-supervised clustering can be formulated as clustering on singly type relational data consisting of attributes and homogeneous relations.

Although this book is mainly focused on the collective clustering framework, it also includes our most recent research on the individual clustering framework, specifically multiple-view relational data clustering, since in some applications, when a large number of types of objects in a relational data set are related to each other in a complicated way, we may want to focus on a certain type of data objects to reduce the model complexity.

Figure 1.1 shows the relations among different areas of relational data clustering. In summary, as a recently booming area, relational data clustering arises in a wide range of applications and is also related to a number of important clustering problems in the literature. Hence, there is a great need for both practical algorithm derivation and theoretical framework construction for relational data clustering, which is the main goal of this book.

1.2   The Content and the Organization of This Book

This book aims at introducing a novel theoretical framework for a new data mining field, relational data clustering, and a family of new algorithms for different relational clustering problems arising in a wide range of important applications.

The organization of this book is as follows. The whole book contains four parts, introduction, models, algorithms, and applications. The introduction part defines the area of relational data clustering and to outline what this book is about; the model part introduces different types of model formulations for relational data clustering; the algorithm part presents various algorithms for the corresponding models; the application part shows applications of the models and algorithms by extensive experimental results. This book focuses on six topics of relational data clustering.

The first topic is clustering on bi-type heterogeneous relational data, in which there are heterogeneous relations between the two types of data objects. For example, a text corpus can be formulated as a bi-type relational data set of documents and words, in which there exist heterogeneous relations between documents and words. Bi-type relational data clustering is also known as co-clustering in the literature. We present a new co-clustering framework, Block Value Decomposition (BVD), for bi-type heterogeneous relational data, which factorizes the relational data matrix into three components: the row-coefficient matrix R, the block value matrix B, and the column-coefficient matrix C. Under this framework, we focus on a special yet very popular case—-nonnegative relational data, and propose a specific novel co-clustering algorithm that iteratively computes the three decomposition matrices based on the multiplicative updating rules.

The second topic is about a more general case of bi-type heterogeneous relational data, multi-type heterogeneous relational data that formulate k-partite graphs with various structures. In fact, many examples of real-world data involve multiple types of data objects that are related to each other, which naturally form k-partite graphs of heterogeneous types of data objects. For example, documents, words, and categories in taxonomy mining, as well as Web pages, search queries, and Web users in a Web search system all form a tri-partite graph; papers, key words, authors, and publication venues in a scientific publication archive form a quart-partite graph. We propose a general model, the relation summary network, to find the hidden structures (the local cluster structures and the global community structures) from a k-partite heterogeneous relation graph. The model provides a principal framework for unsupervised learning on k-partite heterogeneous relation graphs of various structures. Under this model, we derive a novel algorithm to identify the hidden structures of a k-partite heterogeneous relation graph by constructing a relation summary network to approximate the original k-partite heterogeneous relation graph under a broad range of distortion measures.

For the third topic, this book presents homogeneous relational data clustering. In heterogeneous relational data, we have heterogeneous relations between different types of data objects. On the other hand, in homogeneous relational data, there are homogeneous relations between the data objects of a single type. Homogeneous relational data also arise from important applications, such as Web mining, social network analysis, bioinformatics, Very Large-Scale Integration (VLSI) design, and task scheduling. Graph partitioning in the literature can be viewed as a special case of homogeneous relational data clustering. Basically, graph partitioning looks for dense clusters corresponding to strongly intra-connected subgraphs. On the other hand, the goal of homogeneous relational data clustering is more general and challenging. It is to identify both dense clusters and sparse clusters. We propose a general model based on graph approximation to learn relation-pattern-based cluster structures from a graph. The model generalizes the traditional graph partitioning approaches and is applicable to learning various cluster structures. Under this model, we derive a family of algorithms which are flexible to learn various cluster structures and easy to incorporate the prior knowledge of the cluster structures.

The fourth topic is clustering on the most general case of relational data, which contain three types of information: attributes for individual objects, homogeneous relations between objects of the same type, and heterogeneous relations between objects of different types. How to make use of all three types of information to cluster multi-type-related objects simultaneously is a big challenge, since the three types of information have different forms and very different statistical properties. We propose a probabilistic model for relational clustering, which also provides a principal framework to unify various important clustering problems, including traditional attributes-based clustering, semi-supervised clustering, co-clustering, and graph clustering. The proposed model seeks to identify cluster structures for each type of data objects and interaction patterns between different types of objects. Under this model, we propose parametric hard and soft relational clustering algorithms under a large number of exponential family distributions.

The fifth topic is about individual relational clustering framework. On this topic, we propose a general model for multiple-view unsupervised learning. The proposed model introduces the concept of mapping function to make the different patterns from different pattern spaces comparable and hence an optimal pattern can be learned from the multiple patterns of multiple representations. Under this model, we formulate two specific models for two important cases of unsupervised learning: clustering and spectral dimensionality reductions; we derive an iterating algorithm for multiple-view clustering, and a simple algorithm providing a global optimum to multiple spectral dimensionality reduction. We also extend the proposed model and algorithms to evolutionary clustering and unsupervised learning with side information.

The sixth topic is about our most recent research on the evolutionary clustering, which has great potential to incorporate time effects into relational data clustering. Evolutionary clustering is a relatively new research topic in data mining. Evolutionary clustering refers to the scenario where a collection of data evolves over the time; at each time, the collection of the data has a number of clusters; when the collection of the data evolves from one time to another, new data items may join the collection and existing data items may disappear; similarly, new clusters may appear and at the same time existing clusters may disappear. Consequently, both the data items and the clusters of the collection may change over the time, which poses a great challenge to the problem of evolutionary clustering in comparison with the traditional clustering. In this book, we introduce the evolutionary clustering models and algorithms based on Dirichlet processes.

1.3   The Audience of This Book

This book is a monograph on the authors' recent research in relational data clustering, the recently emerging area of data mining and machine learning related to a wide range of applications. Therefore, the expected readership of this book includes all the researchers and system developing engineers working in the areas, including but not limited to, data mining, machine learning, computer vision, multimedia data mining, pattern recognition, statistics, as well as other application areas that use relational data clustering techniques such as Web mining, information retrieval, marketing, and bioinformatics. Since this book is self-contained in the presentations of the materials, this book also serves as an ideal reference book for people who are interested in this new area of relational data clustering. Consequently, in addition, the readership also includes any of those who have this interest or work in a field which needs this reference book. Finally, this book can be used as a reference book for a graduate course on advanced topics of data mining and/or machine learning, as it provides a systematic introduction to this booming new subarea of data mining and machine learning.

1.4   Further Readings

As a newly emerging area of data mining and machine learning, relational data clustering is just in its infant stage; currently there is no dedicated, premier venue for the publications of the research in this area. Consequently, the related work in this area, as the supplementary information to this book for further readings, may be found in the literature of the two parent areas.

In data mining area, related work may be found in the premier conferences such as ACM International Conference on Knowledge Discovery and Data Mining (ACM KDD), IEEE International Conference on Data Mining (IEEE ICDM), and SIAM International Conference on Data Mining (SDM). In particular, related work may be found in the workshop dedicated to the area of relational learning, such as Statistical Relational Learning workshop. For journals, the premier journals in the data mining area may contain related work in relational data clustering, including *IEEE Transactions on Knowledge and Data Engineering* (IEEE TKDE), *ACM Transactions on Data Mining* (ACM TDM), and *Knowledge and Information Systems* (KAIS).

In the machine learning area, related work may be found in the premier conferences such as International Conference on Machine Learning (ICML), Neural Information Processing Systems (NIPS), European Conference on Machine Learning (ECML), European Conference on Principles and Practice of Knowledge Discovery in Databases (PKDD), International Joint Conference on Artificial Intelligence (IJCAI), and Conference on Learning Theory (COLT). For journals, the premier journals in machine learning area may contain related work in relational data clustering, including *Journal of Machine Learning Research* (JMLR) and *Machine Learning Journal* (MLJ).

Part I

Models

Chapter 2

Co-Clustering

A bi-type heterogeneous relational data set consists of two types of data objects with heterogeneous relations between them. Bi-type heterogenous relational data are a very important special case of heterogeneous relational data, since they arise frequently in various important applications. In bi-type heterogeneous relational data clustering, we are interested in clustering two types of data objects simultaneously. This is also known as co-clustering in the literature. In this chapter, we present a new co-clustering framework, Block Value Decomposition (BVD), for bi-type heterogeneous relational data, which factorizes the relational data matrix into three components: the row-coefficient matrix R, the block value matrix B, and the column-coefficient matrix C.

2.1 Introduction

In many applications, such as document clustering, collaborative filtering, and micro-array analysis, the bi-type heterogeneous relational data can be formulated as a two-dimensional matrix representing a set of dyadic data. Dyadic data refer to a domain with two finite sets of objects in which observations are made for *dyads*, i.e., pairs with one element from either set. For the dyadic data in these applications, co-clustering both dimensions of the data matrix simultaneously is often more desirable than traditional one-way clustering. This is due to the fact that co-clustering takes the benefit of exploiting the duality between rows and columns to effectively deal with the high-dimensional and sparse data that are typical in many applications. Moreover, there is an additional benefit for co-clustering to provide both row clusters and column clusters at the same time. For example, we may be interested in simultaneously clustering genes and experimental conditions in bioinformatics applications [29, 31], simultaneously clustering documents and words in text mining [44], and simultaneously clustering users and movies in collaborative filtering.

In this chapter, we propose a new co-clustering framework called Block Value Decomposition (BVD). The key idea is that the latent block structure in a two-dimensional dyadic data matrix can be explored by its triple decomposition. The dyadic data matrix is factorized into three components: the row-coefficient matrix R, the block value matrix B, and the column-coefficient matrix C. The coefficients denote the degrees of the rows and columns associated with their clusters, and the block value matrix is an explicit and compact representation of the hidden block structure of the data matrix.

Under this framework, we develop a specific novel co-clustering algorithm for a special yet very popular case—nonnegative dyadic data that iteratively computes the three decomposition matrices based on the multiplicative updating rules derived from an objective criterion. By intertwining the row clusterings and the column clusterings at each iteration, the algorithm performs an implicitly adaptive dimensionality reduction, which works well for typical high-dimensional and sparse data in many data mining applications. We have proven the correctness of the algorithm by showing that the algorithm is guaranteed to converge and have conducted extensive experimental evaluations to demonstrate the effectiveness and potential of the framework and the algorithm. As compared with the existing co-clustering methods in the literature, the BVD framework as well as the specific algorithm offers an extra capability: it gives an explicit and compact representation of the hidden block structures in the original data which helps understand the interpretability of the data. For example, the block value matrix may be used to interpret the explicit relationship or association between the document clusters and word clusters in a document-word co-clustering.

2.2   Related Work

This work is primarily related to two main areas: co-clustering in data mining and matrix decomposition in matrix computation.

Although most of the clustering literature focuses on one-sided clustering algorithms [5], recently co-clustering has become a topic of extensive interest due to its applications to many problems such as gene expression data analysis [29, 31] and text mining [44]. A representative early work of co-clustering was reported in [71] that identified hierarchical row and column clustering in matrices by a local greedy splitting procedure. The BVD framework proposed in this paper is based on the partitioning-based co-clustering formulation first introduced in [71].

The model-based clustering methods for a two-dimensional data matrix represent another main direction in co-clustering research. These methods (e.g., [66, 67]) typically have clear probabilistic interpretation. However, they are all based on simplistic assumptions on data distributions, such as Gaussian mixture models. There are no such assumptions in the BVD framework.

Recently, information-theory based co-clustering has attracted intensive attention in the literature. The Information Bottleneck (IB) framework [122] was first introduced for one-sided clustering. Later, an agglomerative hard clustering version of the IB method was used in [117] to cluster documents after clustering words. The work in [49] extended the above framework to repeatedly cluster documents and then words. An efficient algorithm was presented in [44] that monotonically increases the preserved mutual information by intertwining both the row and column clusterings at all stages. All these methods suffer from the fundamental limitation for their applications to a co-occurrence matrix since they interpret the data matrix as a joint distribution of two discrete random variables. A more generalized co-clustering framework was presented in [11] wherein any Bregman divergence can be used in the objective function, and various conditional expectation based constraints can be incorporated into the framework.

There have been many research studies that perform clustering based on Singular Value Decomposition (SVD) or eigenvector-based decomposition [28, 38, 47, 113]. The latent semantic indexing method (LSI) [38] projects each data vector into the singular vector space through the SVD, and then conducts the clustering using traditional data clustering algorithms (such as k-means) in the transformed space. The spectral clustering methods based on the graph partitioning theory focus on finding the best cuts of a graph that optimize certain predefined criterion functions. The optimization of the criterion functions usually leads to the computation of singular vectors or eigenvectors of certain graph affinity matrices. Many criterion functions, such as the average cut [28], the average association [113], the normalized cut [113], and the min-max cut [47], have been proposed along with the efficient algorithms for finding the optimal solutions. Since the computed singular vectors or eigenvectors do not correspond directly to individual clusters, the decompositions from SVD- or eigenvector-based methods are difficult to interpret and to map to the final clusters; as a result, traditional data clustering methods such as k-means must be applied in the transformed space.

Recently, another matrix decomposition formulation, Nonnegative Matrix Factorization (NMF) [36], has been used for clustering [133]. NMF has the intuitive interpretation for the result. However, it focuses on one-dimension of the data matrix and does not take advantage of the duality between the rows and the columns of a matrix.

2.3 Model Formulation and Analysis

2.3.1 Block Value Decomposition

We start by reviewing the notion of dyadic data. The notion dyadic refers to a domain with two sets of objects: $\mathcal{X} = \{x_1, \ldots, x_n\}$ and $\mathcal{Y} = \{y_1, \ldots, y_m\}$

in which the observations are made for *dyads* (x, y). Usually a dyad is a scalar value $w(x, y)$, e.g., the frequency of co-occurrence, or the strength of preference/association/expression level. For the scalar *dyads*, the data can always be organized as an n-by-m two-dimensional matrix Z by mapping the row indices into $\mathcal{X}$ and the column indices into $\mathcal{Y}$. Then, each $w(x, y)$ corresponds to one element of Z.

We are interested in simultaneously clustering $\mathcal{X}$ into k disjoint clusters and $\mathcal{Y}$ into l disjoint clusters. Let the k clusters of $\mathcal{X}$ be written as: $\{\hat{x}_1, \ldots, \hat{x}_k\}$, and the l clusters of $\mathcal{Y}$ be written as: $\{\hat{y}_1, \ldots, \hat{y}_l\}$. In other words, we are interested in finding mappings $C_{\mathcal{X}}$ and $C_{\mathcal{Y}}$,

$$C_{\mathcal{X}} : \{x_1, \ldots, x_n\} \rightarrow \{\hat{x}_1, \ldots, \hat{x}_k\}$$
$$C_{\mathcal{Y}} : \{y_1, \ldots, y_n\} \rightarrow \{\hat{y}_1, \ldots, \hat{y}_l\}$$

This is equivalent to finding the block structures of the matrix Z, i.e., finding the $k \times l$ submatrices of Z such that the elements within each submatrix are similar to each other and elements from different submatrices are dissimilar to each other. This equivalence relation can be illustrated by the procedure below.

Suppose that we are given the cluster labels of rows and columns. Let us permute the rows of Z such that the rows within the same cluster are arranged together and the columns within the same cluster are arranged together. Consequently, we have discovered the hidden block structure from the permuted data matrix. On the other hand, if we are given the data matrix with block structure, it is trivial to derive the clustering of rows and columns. The original data matrix and the permuted data matrix in Figure 2.1 give an illustrative example.

Since the elements within each block are similar to each other, we expect one center to represent each block. Therefore a $k \times l$ small matrix is considered as the compact representation for the original data matrix with a $k \times l$ block structure. In the traditional one-way clustering, given the cluster centers and the weights that denote degrees of observations associated with their clusters, one can approximate the original data by linear combinations of the cluster centers. Similarly, we should be able to "reconstruct" the original data matrix by the linear combinations of the block centers. Based on this observation, we formulate the problem of co-clustering dyadic data as the optimization problem of matrix decomposition, i.e., block value decomposition (BVD).

DEFINITION 2.1 *Block value decomposition of a data matrix $Z \in \Re^{n \times m}$ is given by the minimization of*

$$f(R, B, C) = \|Z - RBC\|^2 \tag{2.1}$$

subject to the constraints $\forall ij : R_{ij} \geq 0$ and $C_{ij} \geq 0$, where $\| \cdot \|$ denote Frobenius matrix norm, $R \in \Re^{n \times k}$, $B \in \Re^{k \times l}$, $C \in \Re^{l \times m}$, $k \ll n$, and $l \ll m$.

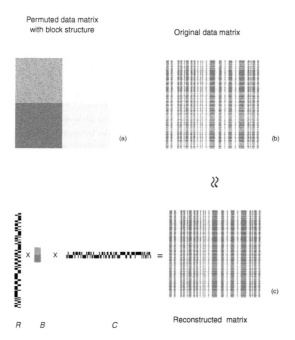

FIGURE 2.1: The original data matrix (b) with a 2×2 block structure which is demonstrated by the permuted data matrix (a). The row-coefficient matrix R, the block value matrix B, and the column-coefficient matrix C give a reconstructed matrix (c) to approximate the original data matrix (b).

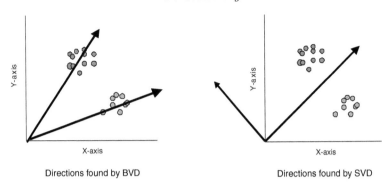

FIGURE 2.2: Illustration of the difference between BVD and SVD.

We call the elements of B as the block values; B as the block value matrix; R as the row-coefficient matrix; and C as the column-coefficient matrix . As is discussed before, B may be considered as a compact representation of Z; R denotes the degrees of rows associated with their clusters; and C denotes the degrees of the columns associated with their clusters. We seek to approximate the original data matrix by the reconstructed matrix, i.e., RBC, as illustrated in Figure 2.1.

Under the BVD framework, the combinations of the components also have an intuitive interpretation. RB is the matrix containing the basis for the column space of Z and BC contains the basis for the row space of Z. For example, for a word-by-document matrix Z, each column of RB captures a base topic of a particular document cluster and each row of BC captures a base topic of a word cluster.

Compared with SVD-based approaches, there are three main differences between BVD and SVD. First, in BVD, it is natural to consider each row or column of a data matrix as an additive combination of the block values since BVD does not allow negative values in R and C. In contrast, since SVD allows the negative values in each component, there is no intuitive interpretation for the negative combinations. Second, unlike the singular vectors in SVD, the basis vectors contained in RB and BC are not necessarily orthogonal. Although singular vectors in SVD have a statistical interpretation as the directions of the variance, they typically do not have clear physical interpretations. In contrast, the directions of the basis vectors in BVD have much more straightforward correspondence to the clusters (Figure 2.2). Third, SVD is a full rank decomposition whereas BVD is a reduced rank approximation. Since the clustering task seeks the reduced or compact representation for the original data, BVD achieves the objective directly, i.e., the final clusters can be easily derived without additional clustering operations. In summary, compared with SVD or eigenvector-based decomposition, the decomposition from BVD has an intuitive interpretation, which is necessary for many data mining applications.

BVD provides a general framework for co-clustering. Depending on different data types in different applications, various formulations and algorithms may be developed under the BVD framework. An interesting observation is that the data matrices in many important applications are typically nonnegative, such as the co-occurrence tables, the performance/rating matrices and the proximity matrices. Some other data may be transformed into the nonnegative form, such as the gene expression data. Therefore, in the rest of the paper, we concentrate on developing a specific novel method under BVD framework, the nonnegative block value decomposition (NBVD).

2.3.2   NBVD Method

In this section we formulate NBVD as an optimization problem and discuss several important properties about NBVD. Discussions in this section can also be applied to general BVD framework.

DEFINITION 2.2   *Nonnegative block value decomposition of a nonnegative data matrix $Z \in \Re^{n \times m}$ (i.e., $\forall ij : Z_{ij} \geq 0$) is given by the minimization of*

$$f(R, B, C) = \|Z - RBC\|^2 \qquad (2.2)$$

subject to the constraints $\forall ij : R_{ij} \geq 0, B_{ij} \geq 0$ and $C_{ij} \geq 0$, where $R \in \Re^{n \times k}$, $B \in \Re^{k \times l}$, $C \in \Re^{l \times m}$, $k \ll n$, and $l \ll m$.

Property 1: The solution to NBVD at a global minimum or a specific local minimum is not unique.

If R, B, and C are a solution to the objective function defined in Equation 2.2 at a global minimum or a specific local minimum, then, $RU, U^{-1}BV$, and $V^{-1}C$ are another solution to f at the same minimum for appropriate positive invertible matrices U and V because

$$(RU)(U^{-1}BV)(V^{-1}C) = RI_k BI_l C = RBC$$

where I_k and I_l are identity matrices.

Therefore, in order to make the solution unique, we may consider normalizing the components. However, the normalization may also change the clustering result. For example, If we normalize R by the column, i.e., let $\tilde{R} = RU$, where $U = diag(R^T e)$ and $e = [1, 1, \ldots, 1]^T$, the cluster labels of some row observations may be changed because the relative weight for association with their row clusters may also be changed. Consequently, whether or not and how to do normalization usually depend on the data and the specific application. For example, in document clustering, typically each document vector is normalized to have unit L^2 norm. Thus, normalizing each column of RB to have unit L^2 norm is desirable, since RB consists of the basis vectors of the document space. Assuming that RB is normalized to RBV, the cluster labels for the documents are given by $V^{-1}C$ instead of C.

Property 2: With the column-normalized R and the row-normalized C, each block value in B may be interpreted as the sum of the elements of the corresponding block.

Let us illustrate the property with an example. Consider a 6×6 word-by-document co-occurrence matrix below:

$$\begin{bmatrix} 0 & 1 & 0 & 6 & 6 & 5 \\ 0 & 0 & 0 & 6 & 6 & 6 \\ 5 & 5 & 5 & 0 & 0 & 0 \\ 5 & 5 & 5 & 0 & 0 & 0 \\ 4 & 0 & 4 & 3 & 3 & 3 \\ 3 & 4 & 4 & 3 & 0 & 3 \end{bmatrix}.$$

Clearly, the matrix may be divided into $3 \times 2 = 6$ blocks. A nonnegative block value decomposition is given as follows:

$$\begin{bmatrix} 0.5 & 0 & 0 \\ 0.5 & 0 & 0 \\ 0 & 0.5 & 0 \\ 0 & 0.5 & 0 \\ 0 & 0 & 0.5 \\ 0 & 0 & 0.5 \end{bmatrix} \times \begin{bmatrix} 1 & 35 \\ 30 & 0 \\ 19 & 15 \end{bmatrix} \times \begin{bmatrix} 0.34 & 0 \\ 0.3 & 0 \\ 0.36 & 0 \\ 0 & 0.36 \\ 0 & 0.3 \\ 0 & 0.34 \end{bmatrix}^T.$$

In the above decomposition, R is normalized by column and C (for the format reason, it is shown as its transpose) is normalized by row and the block value of B is the sum of the elements of the corresponding block. In this example, B can be intuitively interpreted as the co-occurrence matrix of the word clusters and the document clusters. In fact, if we interpret the data matrix as the joint distribution of the words and documents, all the components have a clear probabilistic interpretation. The column-normalized R may be considered as the conditional distribution $p(\text{word}|\text{word cluster})$, the row-normalized C may be considered as the conditional distribution $p(\text{document}|\text{document cluster})$, and B may be considered as the joint distribution of the word clusters and the document clusters.

By property 1, given an NBVD decomposition, we can always define the column-normalized R, the row-normalized C, and the corresponding B as a solution to the original problem. We denote B in this situation as B_s.

Property 3: The variation of the mean block value matrix $\bar{B}$ is a measure of the quality of co-clustering, where $\bar{B}$ is defined as dividing each element of

B_s by the cardinality of the corresponding block.

Intuitively, each element of $\bar{B}$ represents the mean of the corresponding block. Therefore, under the same objective value, the larger the variation of $\bar{B}$, the larger the difference among the blocks, the better separation the co-clustering has.

We propose the simple statistic, the coefficient of variance (CV), to measure the variation $\bar{B}$. Typically, CV is used to measure the variation of the sample from different populations, and thus, it is appropriate to measure the variation of $\bar{B}$ with different dimensions. CV is defined as the ratio between the standard deviation and the mean. Hence, the CV for the mean block value matrix $\bar{B}$ is,

$$CV(\bar{B}) = \frac{\sqrt{\sum_{i,j} (\bar{B}_{ij} - \bar{b})^2/(kl - 1)}}{\bar{b}}, \tag{2.3}$$

where $\bar{b}$ is the mean of $\bar{B}$ and $1 \leq i \leq k$, $1 \leq j \leq l$. To measure the qualities of the row clusters and the column clusters, respectively, we define average CV for the rows of $\bar{B}$, $CV(\bar{B}_r)$, and for the columns of $\bar{B}$, $CV(\bar{B}_c)$, respectively.

$$CV(\bar{B}_r) = \frac{1}{l} \sum_j \frac{\sqrt{\sum_i (\bar{B}_{ij} - \bar{b}_j)^2/(l - 1)}}{\bar{b}_j} \tag{2.4}$$

$$CV(\bar{B}_c) = \frac{1}{k} \sum_i \frac{\sqrt{\sum_j (\bar{B}_{ij} - \bar{b}_i)^2/(l - 1)}}{\bar{b}_i}, \tag{2.5}$$

where $\bar{b}_i$ is the mean of the ith row of $\bar{B}$ and $\bar{b}_j$ is the mean of the jth column of $\bar{B}$. It is necessary to define these statistics in order to provide certain useful information (e.g., to find the optimal number of clusters).

Finally, we compare NBVD with Nonnegative Matrix Factorization (NMF) [36]. Given a nonnegative data matrix V, NMF seeks to find an approximate factorization $V \approx WH$ with non-negative components W and H. Essentially, NMF concentrates on the one-sided, individual clustering and does not take the advantage of the duality between the row clustering and the column clustering. In fact, NMF may be considered as a special case of NBVD in the sense that $WH = WIH$, where I is an identity matrix. By this formulation, NMF is a special case of NBVD and it does co-clustering with the additional restrictions that the number of the row clusters equals to that of the column clusters and that each row cluster is associated with one column cluster. Clearly, NBVD is more flexible to exploit the hidden block structure of the original data matrix than NMF.

Chapter 3

Heterogeneous Relational Data Clustering

In more general cases, heterogeneous relational data consist of more than two types of data objects. Those multiple-type interrelated data objects formulates a k-partite heterogeneous relation graph. The research on mining the hidden structures from a k-partite heterogeneous relation graph is still limited and preliminary. In this chapter, we propose a general model, the relation summary network, to find the hidden structures (the local cluster structures and the global community structures) from a k-partite heterogeneous relation graph. The model provides a principal framework for unsupervised learning on k-partite heterogeneous relation graphs of various structures.

3.1 Introduction

Clustering approaches have traditionally focused on the homogeneous data objects. However, many examples of real-world data involve objects of multiple types that are related to each other, which naturally form k-partite heterogeneous relation graphs of heterogeneous types of nodes. For example, documents, words, and categories in taxonomy mining, as well as Web pages, search queries, and Web users in a Web search system all form a tripartite graph; papers, key words, authors, and publication venues in a scientific publication archive form a quart-partite graph. In such scenarios, using traditional approaches to cluster each type of objects (nodes) individually may not work well due to the following reasons.

First, to apply traditional clustering approaches to each type of data objects individually, the relation information needs to be transformed into feature vectors for each type of objects. In general, this transformation results in high-dimensional and sparse feature vectors, since after the transformation the number of the features for a single type of objects is equal to that of all the objects which are possibly related to this type of objects. For example, if we transform the links between Web pages and Web users as well as search queries into the features for the Web pages, this leads to a huge number of features with sparse values for each Web page. Second, traditional clustering approaches are unable to tackle the interactions among the cluster structures of different types of objects, since they cluster data of a single type based on static features. Note that the interactions could pass along the relations, i.e., there exists influence propagation in a k-partite heterogeneous relation graph. Third, in some data mining applications, users are interested not only in the local cluster structures for each type of objects, but also in the global community structures involving multi-types of objects. For example, in document clustering, in addition to document clusters and word clusters, the relationship between the document clusters and the word clusters is also useful information. It is difficult to discover such global structures by clustering each type of objects individually.

An intuitive attempt to mine the hidden structures from k-partite heterogeneous relation graphs is applying the existing graph partitioning approaches to k-partite heterogeneous relation graphs. This idea may work in some special and simple situations. However, in general, it is infeasible. First, the graph partitioning theory focuses on finding the best cuts of a graph under a certain criterion and it is very difficult to cut different type of relations (links) simultaneously to identify different hidden structures for different types of nodes. Second, by partitioning an entire k-partite heterogeneous relation graph into m subgraphs, one actually assumes that all different types of nodes have the same number of clusters m, which in general is not true. Third, by simply partitioning the entire graph into disjoint subgraphs, the resulting hidden structures are rough. For example, the clusters of different types of nodes are restricted to one-to-one associations.

Therefore, mining hidden structures from k-partite heterogeneous relation graphs has presented a great challenge to traditional clustering approaches. In this chapter, first we propose a general model, the relation summary network, to find the hidden structures (the local cluster structures and the global community structures) from a k-partite heterogeneous relation graph. The basic idea is to construct a new k-partite heterogeneous relation graph with hidden nodes, which "summarize" the link information in the original k-partite heterogeneous relation graph and make the hidden structures explicit, to approximate the original graph. The model provides a principal framework for unsupervised learning on k-partite heterogeneous relation graphs of various structures. Second, under this model, based on the matrix representation of a k-partite heterogeneous relation graph we reformulate the graph approximation as an optimization problem of matrix approximation and derive an iterative algorithm to find the hidden structures from a k-partite heterogeneous relation graph under a broad range of distortion measures. By iteratively updating the cluster structures for each type of nodes, the algorithm takes advantage of the interactions among the cluster structures of different types of nodes and performs an implicit adaptive feature reduction for each type of nodes. Experiments on both synthetic and real data sets demonstrate the promise and effectiveness of the proposed model and algorithm. Third, we also establish the connections between existing clustering approaches and the proposed model to provide a unified view to the clustering approaches.

3.2   Related Work

Graph partitioning on homogeneous graphs has been studied for decades and a number of different approaches, such as spectral approaches [28, 47, 113] and multilevel approaches [25, 63, 75], have been proposed. However, the research on mining cluster structures from k-partite heterogeneous relation graphs of heterogeneous types of nodes is limited. Several noticeable efforts include [43, 69] and [56]. [43, 69] extend the spectral partitioning based on the normalized cut to a bipartite graph. After the deduction, spectral partitioning on the bi-partite graph is converted to a singular value decomposition (SVD). [56] partitions a star-structured k-partite heterogeneous relation graph based on semi-definite programming. In addition to the restriction that they are only applicable to the special cases of k-partite heterogeneous relation graphs, all these algorithms have the restriction that the numbers of the clusters for different types of nodes must be equal and the clusters for different types of objects must have one-to-one associations.

The research on clustering multi-type interrelated objects is also related to this study. Clustering on bi-type interrelated data objects, such as word-document data, is called co-clustering or bi-clustering. Recently, co-clustering has been addressed based on matrix factorization. Both [88] and [85] model the co-clustering as an optimization problem involving a triple matrix factorization. [88] proposes an EM-like algorithm based on multiplicative updating rules and [85] proposes a hard clustering algorithm for binary data. [45] extends the nonnegative matrix factorization to symmetric matrices and shows that it is equivalent to the kernel k-means and the Laplacian-based spectral clustering.

Some efforts on latent variable discovery are also related to co-clustering. PLSA [66] is a method based on a mixture decomposition derived from a latent class model. A two-sided clustering model is proposed for collaborative filtering by [67]. Information-theory based co-clustering has also attracted attention in the literature. [49] extends the information bottleneck (IB) framework [122] to repeatedly cluster documents and then words. [44] proposes a co-clustering algorithm to maximize the mutual information between the clustered random variables subject to the constraints on the number of row and column clusters. A more generalized co-clustering framework is presented by [11] wherein any Bregman divergence can be used in the objective function.

Comparing with co-clustering, clustering on the data consisting of more than two types of data objects has not been well studied in the literature. Several noticeable efforts are discussed as follows. [137] proposes a framework

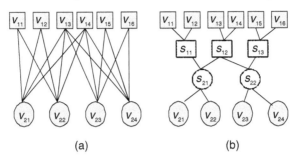

(a)   (b)

FIGURE 3.1: A bipartite graph (a) and its relation summary network (b).

for clustering heterogeneous web objects, under which a layered structure with the link information is used to iteratively project and propagate the cluster results between layers. Similarly, [125] presents an approach named ReCom to improve the cluster quality of interrelated data objects through an iterative reinforcement clustering process. However, there is no sound objective function and theoretical proof on the effectiveness of these algorithms. [87] formulates heterogeneous relational data clustering as a collective factorization on related matrices and derives a spectral algorithm to cluster multi-type interrelated data objects simultaneously. The algorithm iteratively embeds each type of data objects into low-dimensional spaces and benefits from the interactions among the hidden structures of different types of data objects. Recently, a general method based on matrix factorization is independently developed by [115, 116], but had not yet appeared at the time of the write-up

To summarize, unsupervised learning on k-partite heterogeneous relation graphs has been touched from different perspectives due to its high impact in various important applications. Yet, systematic research is still limited. This chapter attempts to derive a theoretically sound general model and algorithm for unsupervised learning on k-partite heterogeneous relation graphs of various structures.

3.3 Relation Summary Network Model

In this section, we derive a general model based on graph approximation to mine the hidden structures from a k-partite heterogeneous relation graph.

Let us start with an illustrative example. Figure 3.1a shows a bipartite graph $G = (V_1, V_2, E)$ where $V_1 = \{v_{11}, \ldots, v_{16}\}$ and $V_2 = \{v_{21}, \ldots, v_{24}\}$ denote two types of nodes and E denotes the edges in G. Even though this graph is simple, it is nontrivial to discover its hidden structures. In Figure

3.1b, we redraw the original graph by adding two sets of new nodes (called hidden nodes), $S_1 = \{s_{11}, s_{12}, s_{13}\}$ and $S_2 = \{s_{21}, s_{22}\}$. Based on the new graph, the cluster structures for each type of nodes are straightforward; V_1 has three clusters: $\{v_{11}, v_{12}\}$, $\{v_{13}, a_{14}\}$, and $\{v_{15}, v_{16}\}$, and V_2 has two clusters, $\{v_{21}, v_{22}\}$ and $\{v_{23}, b_{24}\}$. If we look at the subgraph consisting of only the hidden nodes in Figure 3.1b, we see that it provides a clear skeleton for the global structure of the whole graph, from which it is clear how the clusters of different types of nodes are related to each other; for example, cluster s_{11} is associated with cluster s_{21} and cluster s_{12} is associated with both clusters s_{21} and s_{22}. In other words, by introducing the hidden nodes into the original k-partite heterogeneous relation graph, both the local cluster structures and the global community structures become explicit. Note that if we apply a graph partitioning approach to the bipartite graph in Figure 3.1a to find its hidden structures, no matter how we cut the edges, it is impossible to identify all the cluster structures correctly.

Based on the above observations, we propose a model, the relation summary network (RSN), to mine the hidden structures from a k-partite heterogeneous relation graph. The key idea of RSN is to add a small number of hidden nodes to the original k-partite heterogeneous relation graph to make the hidden structures of the graph explicit. However, given a k-partite heterogeneous relation graph, we are not interested in an arbitrary relation summary network. To ensure a relation summary network to discover the desirable hidden structures of the original graph, we must make RSN as "close" as possible to the original graph. In other words, we aim at an optimal relation summary network, from which we can reconstruct the original graph as precisely as possible. Formally, we define an RSN as follows.

DEFINITION 3.1 *Given a distance function $\mathfrak{D}$, a k-partite heterogeneous relation graph $G = (V_1, \ldots, V_m, E)$, and m positive integers, $k_1, \ldots, k_m$, the relation summary network of G is a k-partite heterogeneous relation graph $G^s = (V_1, \ldots, V_m, S_1, \ldots, S_m, E^s)$, which satisfies the following conditions:*

1. *Each instance node in V_i is adjacent to one and only one hidden node from S_i for $1 \leq i \leq m$ with unit weight;*

2. *$S_i \sim S_j$ in G^s if and only if $V_i \sim V_j$ in G for $i \neq j$ and $1 \leq i, j \leq m$;*

3. *$G^s = \arg\min_F \mathfrak{D}(G, F)$,*

where S_i denotes a set of hidden nodes for V_i and $|S_i| = k_i$ for $1 \leq i \leq m$; $S_i \sim S_j$ denotes that there exist edges between S_i and S_j, and similarly $V_i \sim V_j$; F denotes any k-partite heterogeneous relation graph $(V_1, \ldots, V_m, S_1, \ldots, S_m, E^f)$ satisfying Conditions 1 and 2.

In Definition 3.1, the first condition implies that in an RSN, the instance nodes (the nodes in V_i) are related to each other only through the hidden

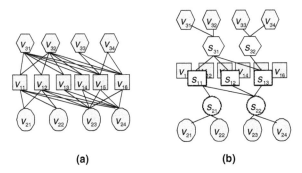

(a)   (b)

FIGURE 3.2: A tripartite graph (a) and its RSN (b)

nodes. Hence, a small number of hidden nodes actually summarize the complex relations (edges) in the original graph to make the hidden structures explicit. Since in this study, our focus is to find disjoint clusters for each type of nodes, the first condition restricts one instance node to be adjacent to only one hidden node with unit weight; however, it is easy to modify this restriction to extend the model to other cases of unsupervised learning on k-partite heterogeneous relation graphs. The second condition implies that if two types of instance nodes V_i and V_j are (or are not) related to each other in the original graph, then the corresponding two types of hidden nodes S_i and S_j in the RSN are (or are not) related to each other. For example, Figure 3.2 shows a tripartite graph and its RSN. In the original graph Figure 3.2a, $V_1 \sim V_2$ and $V_1 \sim V_3$, and hence $S_1 \sim S_2$ and $S_1 \sim S_3$ in its RSN. The third condition states that the RSN is an optimal approximation to the original graph under a certain distortion measure.

Next, we need to define the distance between a k-partite heterogeneous relation graph G and its RSN G^s. Without loss of generality, if $V_i \sim V_j$ in G, we assume that edges between V_i and V_j are complete (if there is no edge between v_{ih} and v_{jl}, we can assume an edge with weight of zero or other special value). Similarly for $S_i \sim S_j$ in G^s. Let $e(v_{ih}, v_{jl})$ denote the weight of the edge (v_{ih}, v_{jl}) in G. Similarly let $e^s(s_{ip}, s_{jq})$ be the weight of the edge (s_{ip}, s_{jq}) in G^s. In the RSN, a pair of instance nodes v_{ih} and v_{jl} are connected through a unique path $(v_{ih}, s_{ip}, s_{jq}, v_{jl})$, in which $e^s(v_{ih}, s_{ip}) = 1$ and $e^s(s_{jq}, v_{jl}) = 1$ according to Definition 3.1. The edge between two hidden nodes (s_{ip}, s_{jq}) can be considered as the "summary relation" between two sets of instance nodes, i.e., the instance nodes connecting with s_{ip} and the instance nodes connecting with s_{jq}. Hence, how good G^s approximates G depends on how good $e^s(s_{ip}, s_{jq})$ approximates $e(v_{ih}, v_{jl})$ for v_{ih} and v_{jl} which satisfy $e^s(v_{ih}, s_{ip}) = 1$ and $e^s(s_{jq}, v_{jl}) = 1$, respectively. Therefore, we define the distance between a k-partite heterogeneous relation graph G and its RSN G^s

as follows:

$$\mathfrak{D}(G, G^s) = \sum_{i,j} \sum_{\substack{V_i \sim V_j, \\ v_{ih} \in V_i, v_{jl} \in V_j, \\ e^s(v_{ih}, s_{ip})=1, \\ e^s(s_{jq}, v_{jl})=1.}} D(e(v_{ih}, v_{jl}), e^s(s_{ip}, s_{jq})), \tag{3.1}$$

where $1 \leq i, j \leq m$, $1 \leq h \leq |V_i|$, $1 \leq l \leq |V_j|$, $1 \leq p \leq |S_i|$, and $1 \leq q \leq |S_j|$.

Let us have an illustrative example. Assume that the edges of the k-partite heterogeneous relation graph in Figure 3.1a have unit weights. If there is no edge between v_{ih} and v_{jl}, we let $e(v_{ih}, v_{jl}) = 0$. Similarly for its RSN in Figure 3.1b. Assume that D is the Euclidean distance function. Hence, based on Equation (3.1), $\mathfrak{D}(G, G^s) = 0$, i.e., from the RSN in Figure 3.1b, we can reconstruct the original graph in Figure 3.1a without any error. For example, the path $(v_{13}, s_{12}, s_{21}, v_{22})$ in the RSN implies that there is an edge between v_{13} and v_{22} in the original graph such that $e(v_{13}, v_{22}) = e^s(s_{12}, s_{21})$. Following this procedure, the original graph can be reconstructed completely.

Note that different definitions of the distances between two graphs lead to different algorithms. In this study, we focus on the definition given in Equation (3.1). One of the advantages of this definition is that it leads to a nice matrix representation for the distance between two graphs, which facilitates to derive the algorithm.

Definition 3.1 and Equation (3.1) provide a general model, the RSN model, to mine the cluster structures for each type of nodes in a k-partite heterogeneous relation graph and the global structures for the whole graph. Compared with the traditional clustering approaches, the RSN model is capable of making use of the interactions (direct or indirect) among the hidden structures of different types of nodes, and through the hidden nodes performing implicit and adaptive feature reduction to overcome the typical high dimensionality and sparsity. Figure 3.3 shows an illustrative example of how the cluster structures of two types of instance nodes affect the similarity between two instance nodes of another type. Suppose that we are to cluster nodes in V_1 (only two nodes in V_1 are shown in Figure 3.3a). Traditional clustering approaches determine the similarity between v_{11} and v_{12} based on their link features, $[1, 0, 1, 0]$ and $[0, 1, 0, 1]$, respectively, and hence, their similarity is inappropriately considered as zero (lowest level). This is a typical situation in a large graph with sparse links. Now suppose that we have derived hidden nodes for V_2 and V_3 as in Figure 3.3b; through the hidden nodes the cluster structures of V_2 change the similarity between v_{11} and v_{12} to 1 (highest level), since the reduced link features for both v_{11} and v_{12} are $[1, 1]$, which is a more reasonable result, since in a sparse k-partite heterogeneous relation graph we expect that two nodes are similar when they are connected to *similar* nodes even though they are not connected to the *same* nodes. If we continue this example, next, v_{11} and v_{12} are connected with the same hidden nodes in S_1 (not shown in the figure); then after the hidden nodes for V_1 are derived, the cluster structures

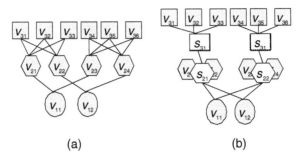

FIGURE 3.3: The cluster structures of V_2 and V_3 affect the similarity between v_{11} and v_{12} through the hidden nodes.

of V_2 and V_3 may be affected in return. In fact, this is the idea of the iterative algorithm to construct an RSN for a k-partite heterogeneous relation graph, which we discuss in the next section.

Chapter 4

Homogeneous Relational Data Clustering

Homogeneous relational data consist of only one type of data objects. In the literature, a special case of homogeneous relational data clustering has been studied as the graph partitioning problem. However, the research on the general case is still limited. In this chapter, we propose a general model based on the graph approximation to learn relation-pattern-based cluster structures from a graph. The model generalizes the traditional graph partitioning approaches and is applicable to learning the various cluster structures.

4.1 Introduction

Learning clusters from homogeneous relational graphs is an important problem in these applications, such as Web mining, social network analysis, bioinformatics, VLSI design, and task scheduling. In many applications, users are interested in strongly intra-connected clusters in which the nodes are intra-cluster close and intercluster loose. Learning this type of the clusters corresponds to finding strongly connected subgraphs from a graph, which has been studied for decades as a graph partitioning problem [28, 77, 113].

In addition to the strongly intra-connected clusters, other types of the clusters also attract an intensive attention in many important applications. For example, in Web mining, we are also interested in the clusters of Web pages that sparsely link to each other but all densely link to the same Web pages [80], such as a cluster of music "fans" Web pages which share the same taste on music and are densely linked to the same set of music Web pages but sparsely linked to each other. Learning this type of the clusters corresponds to finding dense bipartite subgraphs from a graph, which has been listed as one of the five algorithmic challenges in Web search engines [64].

The strongly intra-connected clusters and weakly intra-connected clusters are two basic cluster structures, and various types of clusters can be generated based on them. For example, a Web cluster could take on different structures during its development, i.e., in its early stage, it has the form of bipartite graph, since in this stage the members of the cluster share the same interests

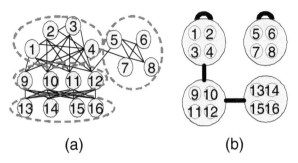

(a)   (b)

FIGURE 4.1: A graph with mixed cluster structures (a) and its cluster prototype graph (b).

(linked to the same Web pages) but have not known (linked to) each other; in the later stage, with members of the cluster start linking to each other, the cluster becomes a hybrid of the aforementioned two basic cluster structures; in the final stage it develops into a larger strongly intra-connected cluster.

These various types of clusters can be unified into a general concept, relation-pattern-based cluster. *A relation-pattern-based cluster is a group of nodes which have the similar relation patterns*, i.e., the nodes within a cluster relate to other nodes in similar ways. Let us have an illustrative example. Figure 4.1a shows a graph of mixed types of clusters. There are four clusters in Figure 4.1a: $C_1 = \{v_1, v_2, v_3, v_4\}$, $C_2 = \{v_5, v_6, v_7, v_8\}$, $C_3 = \{v_9, v_{10}, v_{11}, v_{12}\}$, and $C_4 = \{v_{13}, v_{14}, v_{15}, v_{16}\}$. Within the strongly intra-connected cluster C_1, the nodes have the similar relation patterns, i.e., they all strongly link to the nodes in C_1 (their own cluster) and C_3, and weakly link to the nodes in C_2 and C_4; within the weakly intra-connected cluster C_3, the nodes also have the similar relation patterns, i.e., they all weakly link to the nodes in C_3 (their own cluster), and C_2, strongly link to the nodes in C_1 and C_4; Similarly for the nodes in cluster C_3 and the nodes in cluster C_4. Note that graph partitioning approaches cannot correctly identify the cluster structure of the graph in Figure 4.1a, since they seek only strongly intra-connected clusters by cutting a graph into disjoint subgraphs to minimize edge cuts.

In addition to unsupervised cluster learning applications, the concept of the relation-pattern-based cluster also provides a simple approach for semi-supervised learning on graphs. In many applications, graphs are very sparse and there may exist a large number of isolated or nearly isolated nodes which do not have cluster patterns. However, according to extra supervised information (domain knowledge), these nodes may belong to certain clusters. To incorporate the supervised information, a common approach is to manually label these nodes. However, for a large graph, manually labeling is labor-intensive and expensive. Furthermore, to make use of these labels, instead of supervised learning algorithms, different semi-supervised learning algorithms need to be designed. The concept of the relation-pattern-based cluster provides a simple way to incorporate supervised information by adding virtual nodes to graphs. The idea is that if the nodes belong to the same cluster according to the supervised information, they are linked to the same virtual nodes. Then an algorithm which is able to learn general relation-pattern-based clusters can be directly applied to the graphs with virtual nodes to make use of the supervised information to learn cluster patterns.

For example, to find the hidden classes from a collection of the documents, a common approach is to represent the collection as a graph in which each node denotes a document and each edge weight denotes the similarity between two documents [43,69]. Usually, the similarities are calculated based on the term-frequency vectors of documents. However, there may exist documents which share no or very few words with each other but still belong to the same cluster according to extra domain information. Let us have an illustrative example. In Figure 4.2, the dark color nodes (documents) do not share any words and are not linked to each other. However, they all belong to the "vehicle" cluster. By adding virtual nodes (documents) (light color nodes in Figure 4.2) which are concept documents consisting of popular words for the "vehicle" cluster, the originally isolated document nodes are linked to the virtual document nodes and the supervised information is embedded into the relation patterns.

Therefore, various applications involving unsupervised as well as semi-supervised cluster learning have presented a great need to relation-pattern-based cluster learning algorithms. In this chapter, we propose a general model based on graph approximation to learn the relation-pattern-based cluster structures from a graph. By unifying the traditional edge cut objectives, the model provides a new view to understand the graph partitioning approaches and at the same time it is applicable to learning various cluster structures. Under this model, we derive three novel algorithms to learn the general cluster structures from a graph, which cover three main versions of unsupervised learning algorithms: hard, soft, and balanced versions, to provide a complete family of cluster learning algorithms. This family of algorithms has the following advantages: they are flexible to learn various types of clusters; when

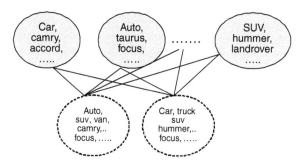

FIGURE 4.2: A graph with virtual nodes.

applied to learning strongly intra-connected clusters, this family evolves to a new family of effective graph partition algorithms; it is easy for the proposed algorithms to incorporate the prior knowledge of the cluster structure into the algorithms. Experimental evaluation and theoretical analysis show the effectiveness and great potential of the proposed model and algorithms.

4.2   Related Work

Graph partitioning divides the nodes of a graph into clusters by finding the best edge cuts of the graph. Several edge cut objectives, such as the average cut [28], average association [113], normalized cut [113], and min-max cut [47], have been proposed. Various spectral algorithms have been developed for these objective functions [28,47,113]. These algorithms use the eigenvectors of a graph affinity matrix, or a matrix derived from the affinity matrix, to partition the graph. Since eigenvectors computed do not correspond directly to individual partitions, a postprocessing approach [136], such as k-means, must be applied to find the final partitions.

Multilevel methods have been used extensively for graph partitioning with the Kernighan-Lin objective, which attempt to minimize the cut in the graph while maintaining equal-sized clusters [25,63,75]. In multilevel algorithms, the graph is repeatedly coarsened level by level until only a small number of nodes are left. Then, an initial partitioning on this small graph is performed. Finally, the graph is uncoarsened level by level, and, at each level, the partitioning from the previous level is refined using a refinement algorithm.

Recently, graph partitioning with an edge cut objective has been shown to be mathematically equivalent to an appropriately weighted kernel k-means objective function [40, 41]. Based on this equivalence, the weighted kernel k-means algorithm has been proposed for graph partitioning [40–42].

Learning clusters from a graph has also been intensively studied in the context of social network analysis [109]. Hierarchical clustering [109,128] has been proposed to learn clusters. Recent algorithms [32,60,94] address several problems related to the prior knowledge of cluster size, the precise definition of inter-nodes similarity measure, and improved computational efficiency [95]. However, their main focus is still learning strongly intra-connected clusters. Some efforts [4,55,65,65,118] can be considered as cluster learning based on stochastic block modeling.

There are efforts in the literature focusing on finding clusters based on dense bipartite graphs [80, 104]. The trawling algorithm [80] extracts clusters (which are called emerging clusters in [80] as the counterpart concept of strongly intra-connected cluster) by first applying the Apriori algorithm to find all possible cores (complete bipartite graphs) and then expanding each core to a full-fledged cluster with the HITS algorithm [79]. [104] proposes a different approach to extract the emerging clusters by finding all bipartite graphs instead of finding cores.

In this chapter, we focus on how to divide the nodes of a homogeneous relational graph into disjoint clusters based on relation patterns.

4.3   Community Learning by Graph Approximation

In this section, we propose a general model to learn relation-pattern-based clusters from a homogeneous relational graph (for convenience, in the rest of the chapter we simply use graph to refer to homogeneous relational graph).

To derive our model to learn latent cluster structure from a graph, we start from the following simpler problem: if the relation-pattern-based cluster structure of any graph is known, can we draw a simple graph with the explicit latent cluster structure (latent relation patterns) to represent the original graph? We present the concept of a *cluster prototype graph* as an answer. A cluster prototype graph consists of a set of the cluster nodes and a set of links, including self-links for individual cluster nodes and inter-links for pairs of cluster nodes.

For example, Figure 4.1b shows a cluster prototype graph for the graph in Figure 4.1a. Note that for convenience, in all the examples, we use 0-1 graphs where the edge weight 0 denotes the absence of an edge between two nodes and the edges with weight 0 are not shown in the graphs. However, all the discussions are applicable to a general weighted graph. In Figure 4.1(b), the top-left cluster node is associated with the nodes of $C_1 = \{v_1, v_2, v_3, v_4\}$ from the original graph; the self-link of the top-left cluster node implies that all its associated nodes are linked to each other; the inter-link between the top-left cluster node and the bottom-left cluster node implies that the nodes of $C_1 = \{v_1, v_2, v_3, v_4\}$ are linked to those of $C_3 = \{v_9, v_{10}, v_{11}, v_{12}\}$. Hence, the cluster prototype graph in Figure 4.1b provides a clear view of the cluster structure and the relation patterns for the original graph in Figure 4.1a. Given the cluster structures of any graph, we can always draw its cluster prototype graph.

Therefore, learning the hidden cluster structures from a graph can be formulated as finding its optimal cluster prototype graph which is the "closest" to the original graph, i.e., based on this cluster prototype graph, the original graph can be constructed most precisely. By representing a graph as an affinity matrix, this problem can be formally formulated as an optimization problem of matrix approximation,

$$\arg\min_{A^*} ||A - A^*||^2, \tag{4.1}$$

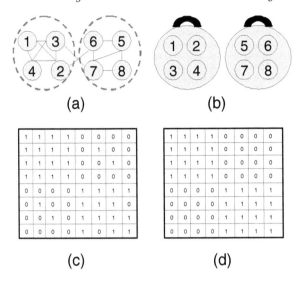

FIGURE 4.3: A graph with strongly connected clusters (a) and its cluster prototype graph (b); the graph affinity matrices for (a) and (b), (c) and (d), respectively.

where $A \in \mathbb{R}_+^{n \times n}$ denotes the affinity matrix of the original graph and $A^* \in \mathbb{R}_+^{n \times n}$ denotes the affinity matrix of a cluster prototype graph. The examples of A and A^* are given in Figure 4.3(c) and 4.3(d), which are the affinity matrices for the original graph Figure 4.3(a) and its cluster prototype graph Figure 4.3(b), respectively.

Due to the special structure of a cluster prototype graph, its affinity matrix can be represented as a product of three factors such that $A^* = CBC^T$, where $C \in \{0,1\}^{n \times k}$ such that k is the number of the clusters and $\sum_j C_{ij} = 1$, i.e., C is an indicator matrix that provides the cluster membership of each node (without loss of generality, we assume that there is no empty cluster); $B \in \mathbb{R}_+^{k \times k}$ such that B is the *cluster structure matrix* that provides an intuitive representation of the cluster structure, since B_{ii} denotes the self-link weight for the ith cluster node and B_{ij} for $i \neq j$ denotes the inter-link weight between the ith and the jth cluster nodes.

Based on the above observation, formally we define the problem of learning clusters from an undirected graph as follows.

DEFINITION 4.1　*Given an undirected graph $G = (\mathcal{V}, \mathcal{E}, A)$ where $A \in \mathbb{R}_+^{n \times n}$ is the affinity matrix, and a positive integer k, the optimized clusters* *are given by the minimization,*

$$\min_{\substack{C\in\{0,1\}^{n\times k},B\in\mathbb{R}_+^{k\times k}\\C\mathbf{1}=\mathbf{1}}} ||A - CBC^T||^2, \tag{4.2}$$

where $\mathbf{1}$ denotes a vector consisting of 1s (we omit its dimension, since it is clear in the context).

Definition 4.1 provides a general model, Cluster Learning by Graph Approximation (CLGA), to learn various cluster structures from graphs. In the CLGA model, the number of the clusters k is given. How to decide the optimal k is a nontrivial model selection problem and beyond the scope of this study.

The problems of finding specific types of clusters can be formulated as special cases of the CLGA model. For example, although there are different formulations of graph partitioning with different objective functions, they all can be considered as special cases of the CLGA model. Since graph partitioning is a very important case of cluster learning, we propose the following theorem to establish the connection between the CLGA model and the existing graph partitioning objectives.

Without loss of generality, we first redefine the cluster indicator matrix C as the following weighted cluster indicator matrix $\tilde{C}$,

$$\tilde{C}_{ij} = \begin{cases} \frac{1}{|\pi_j|^{\frac{1}{2}}} & \text{if } v_i \in \pi_j \\ 0 & \text{otherwise} \end{cases}$$

where $|\pi_j|$ denotes the number of the nodes in the jth cluster. Clearly $\tilde{C}$ still captures the disjoint cluster memberships and $\tilde{C}^T\tilde{C} = I_k$ where I_k denotes a $k \times k$ identity matrix.

THEOREM 4.1
The CLGA model in Definition 4.1 with the extra constraint that B is an identity matrix, i.e.,

$$\min_{\tilde{C}} ||A - \tilde{C}\tilde{C}^T||^2, \tag{4.3}$$

is equivalent to the maximization

$$\max_{\tilde{c}_1,...,\tilde{c}_k} \sum_{p=1}^{k} \tilde{c}_p^T A\tilde{c}_p \tag{4.4}$$

where $\tilde{c}_p$ denotes the pth column vector of $\tilde{C}$.

PROOF Let $\text{tr}(X)$ denote the trace of a matrix X and L denote the objective function in Equation 4.3.

$$L = \text{tr}((A - \tilde{C}\tilde{C}^T)^T(A - \tilde{C}\tilde{C}^T)) \tag{4.5}$$

$$= \operatorname{tr}(A^T A) - 2\operatorname{tr}(\tilde{C}\tilde{C}^T A) + \operatorname{tr}(\tilde{C}\tilde{C}^T \tilde{C}\tilde{C}^T) \qquad (4.6)$$

$$= \operatorname{tr}(A^T A) - 2\operatorname{tr}(\tilde{C}^T A\tilde{C}) + k \qquad (4.7)$$

$$= \operatorname{tr}(A^T A) - 2\sum_{p=1}^{k} \tilde{\mathbf{c}}_p^T A\tilde{\mathbf{c}}_p + k \qquad (4.8)$$

The above deduction uses the property of trace $\operatorname{tr}(XY) = \operatorname{tr}(YX)$. Based on Equation 4.8, since $\operatorname{tr}(A^T A)$ and k, the number of clusters, are constants, the minimization of L is equivalent to the maximization of $\sum_{p=1}^{k} \tilde{\mathbf{c}}_p^T A\tilde{\mathbf{c}}_p$. The proof is completed.   ☐

Theorem 4.1 states that if we fix the cluster structure matrix B as the identity matrix I_k (the more general case is aI_k for any $a \in \mathbb{R}_+$), the CLGA model is reduced to the trace maximization in Equation (4.4). Since various graph partitioning objectives, such as ratio association [113], normalized cut [113], ratio cut [28], and Kernighan-Lin objective [77], can be formulated as the trace maximization [40], Theorem 4.1 establishes the connection between the CLGA model and the existing graph partitioning objectives.

Therefore, the traditional graph partitioning can be considered as a special case of the CLGA model in which B is restricted to be an identity matrix. By fixing B as an identity matrix, the traditional graph partitioning objectives make an implicit assumption about the cluster structure of the target graph, i.e., they assume that the nodes within each cluster are fully connected (the diagonal elements of B are all 1s) and the nodes between clusters are disconnected (the off-diagonal elements of B are all 0s), i.e., the cluster prototype graphs consist of a set of separated cluster nodes with self-links. This assumption is consistent with our intuition about an ideal partitioning (we call this special case of CLGA model Ideal Graph Partitioning (IGP)). However, they cannot catch the cluster structures deviating from the ideal case. For example, for a graph that has one strongly intra-connected subgraph and one relatively weak intra-connected subgraph, assuming B to be $\left[\begin{smallmatrix} 1 & 0 \\ 0 & 0.5 \end{smallmatrix}\right]$ may be better than $\left[\begin{smallmatrix} 1 & 0 \\ 0 & 1 \end{smallmatrix}\right]$. Furthermore, the CLGA model provides the flexibility to learn B under various constraints. If B is relaxed from the identity matrix to any diagonal matrix, i.e., if we assume zero connectivity between clusters but let the algorithm learn the within-cluster connectivity, we obtain a new graph partitioning model as another special case of the general CLGA model, which we call General Graph Partitioning (GGP).

Similarly, with the appropriate constraint on B, the CLGA model may focus on other specific types of cluster structures. For example, by restricting B to be the matrix whose diagonal elements are 0 and off-diagonal elements are 1, CLGA learns the ideal weakly intra-connected clusters among which each pair of clusters forms a dense bipartite graph (call it Ideal Bipartite Cluster Learning (IBCL)); by restricting B to be the matrix whose diagonal elements are 0, CLGA learns clusters of general bipartite subgraphs (call it

TABLE 4.1: A list of variations of the CLGA model

Model	Constraints on B
GCL	no constraints
IGP	Identity matrix
GGP	Diagonal matrix
IBCL	Zero diagonal elements and unit off-diagonal elements
GBCL	Zero diagonal elements

General Bipartite Cluster Learning (GBCL)).

Table 4.1 summarizes several variations of the CLGA model. For simplicity, for the general situation without any constraints on B, we call it General Cluster learning (GCL). The cluster structure matrix B plays an important role in the CLGA model. If we have prior knowledge about the cluster structure of the graph, such as we are only interested in some special types of clusters, it is easy to incorporate it into the model by putting an appropriate constraint on B. If we do not have prior knowledge, we simply use GCL without any constraint on B. For example, if we are only interested in finding the clusters based on the most dense bipartite subgraph, we can simply assume to B be

$$\begin{bmatrix} 0 & 1 & 1 & 0 \\ 1 & 0 & 1 & 0 \\ 1 & 1 & 0 & 0 \\ 0 & 0 & 0 & 0 \end{bmatrix}.$$

Second, although in data mining applications our main goal is the cluster membership matrix C, the cluster structure matrix B could be very useful for interpreting the mining result, since it provides a very intuitive interpretation of the resulting cluster structure.

$$\begin{bmatrix} 0.95 & 0 & 0.9 & 0.01 \\ 0 & 1.0 & 0.01 & 0 \\ 0.9 & 0.01 & 0.98 & 0 \\ 0.01 & 0 & 0 & 0.99 \end{bmatrix}$$

it is easy to know that we obtain four strongly intra-connected clusters, but the first and third clusters could be merged into one cluster to simplify the cluster structure.

Chapter 5

General Relational Data Clustering

In this chapter, we discuss the most general case of the relational data clustering, which makes use of all three types of information: heterogeneous relations altogether, homogeneous relations, and attributes, to cluster multiple-type-related data objects simultaneously. We propose a probabilistic model for relational clustering, which also provides a principal framework to unify various important clustering tasks including traditional attributes-based clustering, semi-supervised clustering, co-clustering, and graph clustering. The proposed model seeks to identify cluster structures for each type of data objects and interaction patterns between different types of objects.

5.1 Introduction

In general, relational data contain three types of information: attributes for individual objects, homogeneous relations between objects of the same type, and heterogeneous relations between objects of different types. For example, for a scientific publication relational data set of papers and authors, the personal information such as affiliation for authors is the attributes; the citation relations among papers are homogeneous relations; the authorship relations between papers and authors are heterogeneous relations. Such data violate the classic IID assumption in machine learning and statistics and present huge challenges to traditional clustering approaches. An intuitive solution is that we transform relational data into flat data and then cluster each type of objects independently. However, this may not work well due to the following reasons.

First, the transformation causes the loss of relation and structure information [48]. Second, the traditional clustering approaches are unable to tackle the influence propagation in clustering relational data, i.e., the hidden patterns of different types of objects could affect each other both directly and indirectly (pass along relation chains). Third, in some data mining applications, users are interested not only in the hidden structures for each type of objects, but also in interaction patterns involving multi-types of objects. For example, in document clustering, in addition to document clusters and word clusters, the relationship between document clusters and word clusters is also useful information. It is difficult to discover such interaction patterns by clustering each type of objects individually.

Moreover, a number of important clustering problems, which have been of intensive interest in the literature, can be viewed as special cases of relational clustering. For example, graph clustering (partitioning) [25, 28, 47, 63, 75, 113] can be viewed as clustering on single-type relational data consisting of only homogeneous relations (represented as a graph affinity matrix); co-clustering [11, 44] which arises in important applications such as document clustering and micro-array data clustering can be formulated as clustering on bi-type relational data consisting of only heterogeneous relations. Recently, semi-supervised clustering [14, 124] has attracted a significant attention, which is a special type of clustering using both labeled and unlabeled data. In Section 11.3, we show that semi-supervised clustering can be formulated as clustering on single-type relational data consisting of attributes and homogeneous relations.

Therefore, relational data present not only huge challenges to traditional unsupervised clustering approaches, but also great need for theoretical unification of various clustering tasks. In this chapter, we propose a probabilistic model for relational clustering, which also provides a principal framework to unify various important clustering tasks, including traditional attributes-based clustering, semi-supervised clustering, co-clustering, and graph clustering. The proposed model seeks to identify cluster structures for each type of data objects and interaction patterns between different types of objects. It is applicable to relational data of various structures. Under this model, we propose parametric hard and soft relational clustering algorithms under a large number of exponential family distributions. The algorithms are applicable to various relational data from various applications and at the same time unify a number of stat-of-the-art clustering algorithms: co-clustering algorithms, the k-partite graph clustering, Bregman k-means, and semi-supervised clustering based on hidden Markov random fields.

5.2   Related Work

Clustering on a special case of relational data, bi-type relational data consisting of only heterogeneous relations, such as the word-document data, is called co-clustering or bi-clustering. Several previous efforts related to co-clustering are model based [66, 67]. Spectral graph partitioning has also been applied to bi-type relational data [43, 69]. These algorithms formulate the data matrix as a bipartite graph and seek to find the optimal normalized cut for the graph. Due to the nature of a bipartite graph, these algorithms have the restriction that the clusters from different types of objects must have one-to-one associations. Information-theory based co-clustering has also attracted attention in the literature. [44] proposes a co-clustering algorithm to maximize the mutual information between the clustered random variables subject to the constraints on the number of row and column clusters. A more generalized co-clustering framework is presented in [11] wherein any Bregman divergence can be used in the objective function. Recently, co-clustering has been addressed based on matrix factorization. [88] proposes an EM-like algorithm based on multiplicative updating rules.

Graph clustering (partitioning) clusters homogeneous data objects based on pairwise similarities, which can be viewed as homogeneous relations. Graph partitioning has been studied for decades and a number of different approaches, such as spectral approaches [28, 47, 113] and multilevel approaches [25, 63, 75], have been proposed. Some efforts [4, 55, 65, 65, 118] based on stochastic block modeling also focus on homogeneous relations.

Compared with co-clustering and homogeneous-relation-based clustering, clustering on general relational data, which may consist of more than two types of data objects with various structures, has not been well studied in the literature. Several noticeable efforts are discussed as follows. [57, 120] extend the probabilistic relational model to the clustering scenario by introducing latent variables into the model; these models focus on using attribute information for clustering. [56] formulates star-structured relational data as a star-structured m-partite graph and develops an algorithm based on semi-definite programming to partition the graph. [86] formulates multi-type relational data as k-partite graphs and proposes a family of algorithms to identify the hidden structures of a k-partite graph by constructing a relation summary network to approximate the original k-partite graph under a broad range of distortion measures. The above graph-based algorithms do not consider attribute information.

Some efforts on relational clustering are based on inductive logic programming [68, 78, 102]. Based on the idea of mutual reinforcement clustering, [137] proposes a framework for clustering heterogeneous Web objects and [125] presents an approach to improve the cluster quality of interrelated data objects through an iterative reinforcement clustering process. There are no sound objective function and theoretical proof on the effectiveness and correctness (convergence) of the mutual reinforcement clustering. Some efforts [18, 73, 134, 135] in the literature focus on how to measure the similarities or choosing cross-relational attributes.

To summarize, the research on relational data clustering has attracted substantial attention, especially in the special cases of relational data. However, there is still limited and preliminary work on general relational data clustering.

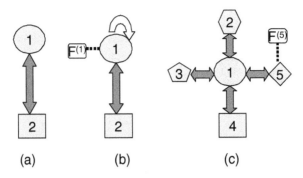

FIGURE 5.1: Examples of the structures of relational data.

5.3   Mixed Membership Relational Clustering

With different compositions of three types of information: attributes, homogeneous relations, and heterogeneous relations, relational data could have very different structures. Figure 5.1 shows three examples of the structures of relational data. Figure 5.1a refers to simple bi-type relational data with only heterogeneous relations such as the word-document data. Figure 5.1b represents a bi-type data with all types of information, such as actor-movie data, in which actors (type 1) have attributes such as gender; actors are related to each other by collaboration in movies (homogeneous relations); actors are related to movies (type 2) by taking roles in movies (heterogeneous relations). Figure 5.1c represents the data consisting of companies, customers, suppliers, shareholders, and advertisement media in which customers (type 5) have attributes.

In this chapter, we represent a relational data set as a set of matrices. Assume that a relational data set has m different types of data objects, $\mathcal{X}^{(1)} = \{x_i^{(1)}\}_{i=1}^{n_1}, \ldots, \mathcal{X}^{(m)} = \{x_i^{(m)}\}_{i=1}^{n_m}$, where n_j denotes the number of objects of the jth type and $x_p^{(j)}$ denotes the name of the pth object of the jth type. We represent the observations of the relational data as three sets of matrices: attribute matrices $\{F^{(j)} \in \mathbb{R}^{d_j \times n_j}\}_{j=1}^m$, where d_j denotes the dimension of the attributes for the jth type objects and $F_{\cdot p}^{(j)}$ denotes the attribute vector for object $x_p^{(j)}$; homogeneous relation matrices $\{S^{(j)} \in \mathbb{R}^{n_j \times n_j}\}_{j=1}^m$, where $S_{pq}^{(j)}$ denotes the relation between $x_p^{(j)}$ and $x_q^{(j)}$; heterogeneous relation matrices $\{R^{(ij)} \in \mathbb{R}^{n_i \times n_j}\}_{i,j=1}^m$, where $R_{pq}^{(ij)}$ denotes the relation between $x_p^{(i)}$ and $x_q^{(j)}$. The above representation is a general formulation. In real applications, not every type of objects necessarily has attributes, homogeneous relations, and heterogeneous relations. For example, the relational data set in Figure 5.1a is represented by only one heterogeneous matrix $R^{(12)}$, and the one in Figure 5.1b is represented by three matrices: $F^{(1)}$, $S^{(1)}$, and $R^{(12)}$. Moreover, for a specific clustering task, we may not use all available attributes and relations after feature or relation selection preprocessing.

Mixed membership models, which assume that each object has mixed membership denoting its association with classes, have been widely used in the applications involving soft classification [50], such as matching words and pictures [106], race genetic structures [106, 129], and classifying scientific publications [51].

In this chapter, we propose a relational mixed membership model to cluster relational data (we refer to the model as *mixed membership relational clustering* or MMRC throughout the rest of the chapter).

Assume that each type of objects $\mathcal{X}^{(j)}$ has k_j latent classes. We represent the membership vectors for all the objects in $\mathcal{X}^{(j)}$ as a membership matrix $\Lambda^{(j)} \in [0,1]^{k_j \times n_j}$ such that the sum of elements of each column $\Lambda^{(j)}_{\cdot p}$ is 1 and $\Lambda^{(j)}_{\cdot p}$ denotes the membership vector for object $x^{(j)}_p$, i.e., $\Lambda^{(j)}_{gp}$ denotes the probability that object $x^{(j)}_p$ associates with the gth latent class. We also write the parameters of the distributions to generate attributes, homogeneous relations, and heterogeneous relations in matrix forms. Let $\Theta^{(j)} \in \mathbb{R}^{d_j \times k_j}$ denote the distribution parameter matrix for generating attributes $F^{(j)}$ such that $\Theta^{(j)}_{\cdot g}$ denotes the parameter vector associated with the gth latent class. Similarly, $\Gamma^{(j)} \in \mathbb{R}^{k_j \times k_j}$ denotes the parameter matrix for generating homogeneous relations $S^{(j)}$; $\Upsilon^{(ij)} \in \mathbb{R}^{k_i \times k_j}$ denotes the parameter matrix for generating heterogeneous relations $R^{(ij)}$. In summary, the parameters of the MMRC model are

$$\Omega = \{\{\Lambda^{(j)}\}^m_{j=1}, \{\Theta^{(j)}\}^m_{j=1}, \{\Gamma^{(j)}\}^m_{j=1}, \{\Upsilon^{(ij)}\}^m_{i,j=1}\}.$$

In general, the meanings of the parameters, Θ, Λ, and Υ, depend on the specific distribution assumptions. However, in Section 11.1.1, we show that for a large number of exponential family distributions, these parameters can be formulated as expectations with intuitive interpretations.

Next, we introduce the latent variables into the model. For each object x^j_p, a latent cluster indicator vector is generated based on its membership parameter $\Lambda^{(j)}_{\cdot p}$, which is denoted as $C^{(j)}_{\cdot p}$, i.e., $C^{(j)} \in \{0,1\}^{k_j \times n_j}$ is a latent indicator matrix for all the jth type objects in $\mathcal{X}^{(j)}$.

Finally, we present the generative process of the observations, $\{F^{(j)}\}^m_{j=1}$, $\{S^{(j)}\}^m_{j=1}$, and $\{R^{(ij)}\}^m_{i,j=1}$ as follows:

1. For each object $x^{(j)}_p$

    - Sample $C^{(j)}_{\cdot p} \sim Multinomial(\Lambda^{(j)}_{\cdot p}, 1)$.

2. For each object $x^{(j)}_p$

    - Sample $F^{(j)}_{\cdot p} \sim Pr(F^{(j)}_{\cdot p} | \Theta^{(j)} C^{(j)}_{\cdot p})$.

3. For each pair of objects $x_p^{(j)}$ and $x_q^{(j)}$

- Sample $S_{pq}^{(j)} \sim Pr(S_{pq}^{(j)} | (C_{\cdot p}^{(j)})^T \Gamma^{(j)} C_{\cdot q}^{(j)})$.

4. For each pair of objects $x_p^{(i)}$ and $x_q^{(j)}$

- Sample $R_{pq}^{(ij)} \sim Pr(R_{pq}^{(ij)} | (C_{\cdot p}^{(i)})^T \Upsilon^{(ij)} C_{\cdot q}^{(j)})$.

In the above generative process, a latent indicator vector for each object is generated based on multinomial distribution with the membership vector as parameters. Observations are generated independently conditioning on latent indicator variables. The parameters of condition distributions are formulated as products of the parameter matrices and latent indicators, i.e.,

$$Pr(F_{\cdot p}^{(j)} | C_{\cdot p}^{(j)}, \Theta^{(j)}) = Pr(F_{\cdot p}^{(j)} | \Theta^{(j)} C_{\cdot p}^{(j)}),$$

$$Pr(S_{pq}^{(j)} | C_{\cdot p}^{(j)}, C_{\cdot q}^{(j)}, \Gamma^{(j)}) = Pr(S_{pq}^{(j)} | (C_{\cdot p}^{(j)})^T \Gamma^{(j)} C_{\cdot q}^{(j)})$$

and

$$Pr(R_{pq}^{(ij)} | C_{\cdot p}^{(i)}, C_{\cdot q}^{(j)}, \Upsilon^{(ij)}) = Pr(R_{pq}^{(ij)} | (C_{\cdot p}^{(i)})^T \Upsilon^{(ij)} C_{\cdot q}^{(j)}).$$

Under this formulation, an observation is sampled from the distributions of its associated latent classes. For example, if $C_{\cdot p}^{(i)}$ indicates that $x_p^{(i)}$ is with the gth latent class and $C_{\cdot q}^{(j)}$ indicates that $x_q^{(j)}$ is with the hth latent class, then $(C_{\cdot p}^{(i)})^T \Upsilon^{(ij)} C_{\cdot q}^{(j)} = \Upsilon_{gh}^{(ij)}$. Hence, we have $Pr(R_{pq}^{(ij)} | \Upsilon_{gh}^{(ij)})$ implying that the relation between $x_p^{(i)}$ and $x_q^{(j)}$ is sampled by using the parameter $\Upsilon_{gh}^{(ij)}$.

With matrix representation, the joint probability distribution over the observations and the latent variables can be formulated as follows:

$$Pr(\Psi | \Omega) = \prod_{j=1}^{m} Pr(C^{(j)} | \Lambda^{(j)}) \prod_{j=1}^{m} Pr(F^{(j)} | \Theta^{(j)} C^{(j)})$$

$$\prod_{j=1}^{m} Pr(S^{(j)} | (C^{(j)})^T \Gamma^{(j)} C^{(j)}) \prod_{i=1}^{m} \prod_{j=1}^{m} Pr(R^{(ij)} | (C^{(i)})^T \Upsilon^{(ij)} C^{(j)}),$$

$$(5.1)$$

where $\Psi = \{ \{C^{(j)}\}_{j=1}^{m}, \{F^{(j)}\}_{j=1}^{m}, \{S^{(j)}\}_{j=1}^{m}, \{R^{(ij)}\}_{i,j=1}^{m} \}, Pr(C^{(j)} | \Lambda^{(j)}) = \prod_{p=1}^{n_j}$ $Multinomial(\Lambda_{\cdot p}^{(j)}, 1), Pr(F^{(j)} | \Theta^{(j)} C^{(j)}) = \prod_{p=1}^{n_j} Pr(F_{\cdot p}^{(j)} | \Theta^{(j)} C_{\cdot p}^{(j)}),$ $Pr(S^{(j)} | (C^{(j)})^T \Gamma^{(j)} C^{(j)}) = \prod_{p,q=1}^{n_j} Pr(S_{pq}^{(j)} | (C_{\cdot p}^{(j)})^T \Gamma^{(j)} C_{\cdot q}^{(j)}),$ and similarly for $R^{(ij)}$.

5.4 Spectral Relational Clustering

In this section, we propose a general spectral clustering model for general relational data based on factorizing multiple related matrices.

Given m sets of data objects, $\mathcal{X}_1 = \{x_{11}, \ldots, x_{1n_1}\}, \ldots, \mathcal{X}_m = \{x_{m1}, \ldots, x_{mn_m}\}$, which refer to m different types of objects relating to each other, we are interested in simultaneously clustering $\mathcal{X}_1$ into k_1 disjoint clusters, $\ldots$, and $\mathcal{X}_m$ into k_m disjoint clusters. We call this task as *collective clustering on heterogeneous relational data* .

To derive a general model for collective clustering, we first formulate relational data as a set of related matrices in which two matrices are related in the sense that their row indices or column indices refer to the same set of objects. First, if there exist relations between $\mathcal{X}_i$ and $\mathcal{X}_j$ (denoted as $\mathcal{X}_i \sim \mathcal{X}_j$), we represent them as a relation matrix $R^{(ij)} \in \mathbb{R}^{n_i \times n_j}$, where an element $R_{pq}^{(ij)}$ denotes the relation between x_{ip} and x_{jq}. Second, a set of objects $\mathcal{X}_i$ may have its own features, which could be denoted by a feature matrix $F^{(i)} \in \mathbb{R}^{n_i \times f_i}$, where an element $F_{pq}^{(i)}$ denotes the qth feature values for the object x_{ip} and f_i is the number of features for $\mathcal{X}_i$.

It has been shown that the hidden structure of a data matrix can be explored by its factorization [36, 88]. Motivated by this observation, we propose a general model for collective clustering, which is based on factorizing the multiple related matrices. In relational data, the cluster structure for a type of objects $\mathcal{X}_i$ may be embedded in multiple related matrices; hence it can be exploited in multiple related factorizations. First, if $\mathcal{X}_i \sim \mathcal{X}_j$, then the cluster structures of both $\mathcal{X}_i$ and $\mathcal{X}_j$ are reflected in the triple factorization of their relation matrix $R^{(ij)}$ such that $R^{(ij)} \approx C^{(i)} A^{(ij)} (C^{(j)})^T$ [88], where $C^{(i)} \in \{0,1\}^{n_i \times k_i}$ is a *cluster indicator matrix* for $\mathcal{X}_i$ such that $\sum_{q=1}^{k_i} C_{pq}^{(i)} = 1$ and $C_{pq}^{(i)} = 1$ denotes that the pth object in $\mathcal{X}_i$ is associated with the qth cluster. Similarly $C^{(j)} \in \{0,1\}^{n_j \times k_j}$. $A^{(ij)} \in \mathbb{R}^{k_i \times k_j}$ is the *cluster association matrix* such that A_{pq}^{ij} denotes the association between cluster p of $\mathcal{X}_i$ and cluster q of $\mathcal{X}_j$. Second, if $\mathcal{X}_i$ has a feature matrix $F^{(i)} \in \mathbb{R}^{n_i \times f_i}$, the cluster structure is reflected in the factorization of $F^{(i)}$ such that $F^{(i)} \approx C^{(i)} B^{(i)}$, where $C^{(i)} \in \{0,1\}^{n_i \times k_i}$ is a cluster indicator matrix, and $B^{(i)} \in \mathbb{R}^{k_i \times f_i}$ is the feature basis matrix which consists of k_i basis (cluster center) vectors in the feature space.

Based on the above discussions, formally we formulate the task of collective clustering on relational data as the following optimization problem. Considering the most general case, we assume that in relational data, every pair of $\mathcal{X}_i$ and $\mathcal{X}_j$ is related to each other and every $\mathcal{X}_i$ has a feature matrix $F^{(i)}$.

DEFINITION 5.1 *Given m positive numbers $\{k_i\}_{1 \leq i \leq m}$ and relational* *data* $\{\mathcal{X}_1, \ldots, \mathcal{X}_m\}$, *which is described by a set of relation matrices* $\{R^{(ij)} \in \mathbb{R}^{n_i \times n_j}\}_{1 \leq i < j \leq m}$, *a set of feature matrices* $\{F^{(i)} \in \mathbb{R}^{n_i \times f_i}\}_{1 \leq i \leq m}$, *as well as a set of weights* $w_a^{(ij)}, w_b^{(i)} \in R_+$ *for different types of relations and features, the task of the collective clustering on the relational data is to minimize*

$$L = \sum_{1 \leq i < j \leq m} w_a^{(ij)} ||R^{(ij)} - C^{(i)} A^{(ij)} (C^{(j)})^T||^2$$
$$+ \sum_{1 \leq i \leq m} w_b^{(i)} ||F^{(i)} - C^{(i)} B^{(i)}||^2 \qquad (5.2)$$

w.r.t. $C^{(i)} \in \{0,1\}^{n_i \times k_i}$, $A^{(ij)} \in \mathbb{R}^{k_i \times k_j}$, *and* $B^{(i)} \in \mathbb{R}^{k_i \times f_i}$ *subject to the constraints:* $\sum_{q=1}^{k_i} C_{pq}^{(i)} = 1$, *where* $1 \leq p \leq n_i$, $1 \leq i < j \leq m$, *and* $|| \cdot ||$ *denotes the Frobenius norm for a matrix.*

We call the model proposed in Definition 5.1 as the Collective Factorization on Related Matrices (CFRM).

The CFRM model clusters multi-type interrelated data objects simultaneously based on both relation and feature information. The model exploits the interactions between the hidden structures of different types of objects through the related factorizations which share matrix factors, i.e., cluster indicator matrices. Hence, the interactions between hidden structures work in two ways. First, if $\mathcal{X}_i \sim \mathcal{X}_j$, the interactions are reflected as the duality of row clustering and column clustering in $R^{(ij)}$. Second, if two types of objects are indirectly related, the interactions pass along the relation "chains" by a chain of related factorizations, i.e., the model is capable of dealing with influence propagation. In addition to local cluster structure for each type of objects, the model also provides the global structure information by the cluster association matrices, which represent the relations among the clusters of different types of objects.

Chapter 6

Multiple-View Relational Data Clustering

In some applications, when a large number of types of objects in a relational data set are related to each other in a complicated way, we may want to focus on a certain type of data objects to reduce the model complexity. In this situation, the same type of data objects is described by different sets of features and different sets of relations. Hence, the relational data clustering problem can be viewed as a case of different view learning in this situation. In this chapter, we propose a general model for multiple-view unsupervised learning. The proposed model introduces the concept of mapping function to make the different patterns from different pattern spaces comparable and hence an optimal pattern can be learned from the multiple patterns of multiple representations.

6.1 Introduction

In many important data mining applications, the same instances have multiple representations from different spaces (views). These multiple representations could be from different feature vector spaces; for example, in bioinformatics, genes can be represented in the expression vector space (corresponding to the genetic activity under the biological conditions) and also in the term vector space (corresponding to the text information related to the genes) [61]. Multiple representations could also be from different relation spaces; for example, in social network analysis, the same instances could be related to each other in multiple networks (graphs), such as e-mail networks, collaboration networks, and organization hierarchy. Finally, multiple presentations could be from mixed views of vector spaces and relation spaces; for example, Web pages can have following multiple representations: the term vector corresponding to words occurring in the pages themselves, the hyperlinks between the pages, and the term vectors corresponding to words contained in anchor text of links pointing to the pages.

Multiple-view data raise a natural, yet nonstandard new problem: how to learn a consensus pattern based on multiple representations, which is more accurate and robust than patterns based on a single view. Due to the phenomenal impact of multiple-view data on many applications, multiple-view learning is attracting more and more attentions [103]. In this work, we consider the problem of multiple-view unsupervised learning, i.e., mining hidden patterns, such as clustering and low-dimension embedding, from multiple-view data without the supervised information (such as class labels).

There are two directions for seeking solutions to multiple-view unsupervised learning. One is to design centralized algorithms, which make use of multiple representations simultaneously to mine hidden patterns from the data. The top challenge for the centralized algorithms is the diversity of the multiple representations, i.e., different representations not only could be of different formulations (vectors or graphs), but also could have very different statistical properties (continuous values with various distributions or discrete values with various distributions). Let us consider the scenario of clustering. A large number of clustering algorithms have been proposed in the literature for various types of data with various statistical properties [16, 72]. For example, Gaussian EM algorithm is a classic algorithm for data with spherical shape clusters; subspace clustering approaches are designed for high-dimensional data; spectral graph clustering approaches are popular for weighted undirected graphs. Hence, in many situations, designing a centralized multiple-view clustering algorithm is close to designing a single algorithm to accomplish several different difficult tasks which are originally handled by several different clustering algorithms. This direction is so difficult that the existing efforts usually have to restrict themselves to special cases of multiple-view data with strong assumptions. For example, [19] focuses on the data of two sets of features, which are assumed to be independent of each other and can be handled by the same algorithm such as k-means or multinomial EM; [139] deals with the data of two graphs which all are assumed to be well explained by a mixture of Markov chains. The above examples also imply that for different types of multiple-view data, we need to design different centralized algorithms.

Another direction to solve multiple-view unsupervised clustering is the distributed approaches, which have rarely been addressed in the literature of multiple-view unsupervised learning. The idea of the distributed approaches is to learn hidden patterns individually from each representation of the multiple-view data and then learn the optimal hidden patterns from those multiple patterns. Compared with the centralized approaches, the distributed approaches have the following advantages. First, they take base unsupervised learning methods for different representations as a black box, allowing practitioners to select the most appropriate method for each representation. Second, under a distributed framework, an algorithm which learns an optimal pattern from multiple patterns can be used for various types of multiple-view data, since it does not work on the data directly. Third, for the multiple-view data under the privacy-preserving scenario, a distributed algorithm is still a feasible solution while a centralized algorithm is not. For example, for a set of customers of different companies, one representation is from a bank and another is from a credit union; due to the privacy issue, the companies cannot share the information (representations), though they want to use both representations to learn hidden patterns for the customers. In such a scenario, the distributed approach provides a natural solution, since the companies could just mine patterns based on their own data and share the patterns (such as clusterings) instead of the original data.

Under a distributed framework, choosing the most appropriate unsupervised clustering for each representation is left for domain experts. Hence, the key problem is how to learn the optimal pattern from multiple patterns. The main challenge of the problem is that since different patterns from different representations exist in different spaces, they cannot be compared directly.

In this chapter, we address the problem of the multiple-view unsupervised learning under a distributed framework. Our main contributions can be summarized as follows:

1. We propose a general model for multiple-view unsupervised learning, which is applicable to various types of unsupervised learning on various types of multiple-view data. We show how to formulate the problems of multiple-view clustering and multiple-view spectral embedding (dimensionality reduction).

2. We derive an iterating algorithm to solve the constrained non-convex problem of multiple-view clustering, which iteratively updates the optimal clustering pattern matrix and the mapping matrices until convergence. We prove that the objective function is nonincreasing under the updating rules and hence the convergence of the algorithm is guaranteed.

3. We derive a simple algorithm for multiple spectral embedding, which provides a global optimal solution to the problem (Section 3.2).

4. We also introduce extensions to our algorithms to handle two important learning settings: unsupervised learning with side information and evolutionary clustering.

6.2 Related Work

Multiple-view learning was introduced by [23] and [74] in the semi-supervised setting. They propose the co-training approach to train a classifier from two representations with both labeled and unlabeled instances. The idea of the co-training approach is to train one learner on each view of the labeled instances and then to iteratively let each learner label the unlabeled instances it predicts with the highest confidence. Given the independence between the learners, newly labeled examples from one learner may give the other learner new information to improve its model. [34] extends co-training to explicitly measure the degree of agreement between the rules in different views. [35] proposes PAC bounds for co-training. [2] explains the co-training based on Bootstrapping theory by showing that given a certain type of independence between the learners, the disagreement between two learners gives an upper bound on the error rate; unlabeled data can be used to minimize the disagreement between the learners, and hence improve their combined accuracy. A number of approaches have followed and extended the original co-training idea [24, 59, 99, 100].

Multiple-view unsupervised learning is a fairly new topic. In fact, the subfields such as multiple-view dimensionality reduction has not been touched in the literature, though there is limited work on multiple-view clustering [19, 37, 139]. As discussed in Section 6.1, these approaches are centralized and focus on the simple case of two views with strong assumptions. [19] proposes clustering algorithms for the multiple-view data with two independent views, which can be considered as an extension to the purely unsupervised setting. [37] also assumes two independent views for a multiple view data set and proposes a spectral clustering algorithm which creates a bipartite graph and is based on the minimizing-disagreement idea. [139] proposes a spectral clustering algorithm for multiple graphs, which generalizes the normalized cut from a single view to that from multiple views.

Ensemble clustering [54, 89, 119, 123] is a related field, though its focus is on how to generate and combine different clusterings for a Single-View-Data set. Clustering multi-type objects is also close to multiple-view clustering [56, 86, 120, 125]. [120] extends the probabilistic relational model to clustering multi-type relational data by introducing latent variables into the model; these models focus on using attribute information for clustering. [56] formulates star-structured relational data as a star-structured m-partite graph and develops an algorithm based on semi-definite programming to partition the graph. [125] presents an approach to improve the cluster quality of multitype interrelated data objects through an iterative reinforcement clustering process.

6.3   Background and Model Formulation

No matter in unsupervised learning or supervised learning, we have input data $\{\mathbf{x}_1, \mathbf{x}_2, \ldots, \mathbf{x}_n\}$. However, in unsupervised learning, we have no supervised target output, which is given in supervised learning, such as the class labels given in a classification task. The goal of unsupervised learning is to build "concise representations" (patterns) from the data that can be used for reasoning, decision making, and communications, etc. Two classic examples of unsupervised learning are clustering and dimensionality reduction.

If we represent an input data set $\{\mathbf{x}_1, \mathbf{x}_2, \ldots, \mathbf{x}_n\}$ for unsupervised learning as a matrix $X \in \mathbb{R}^{n \times d}$, then in general the learned pattern can also be formulated as a matrix $A \in \mathcal{A}$ (we call A *pattern matrix* and $\mathcal{A}$ *pattern space*), where the pattern space $\mathcal{A}$ has a lower dimension than the original data space, i.e., $\mathcal{A} \subseteq \mathbb{R}^{n \times k}$ with $k \ll d$. For different types of unsupervised learning, the pattern spaces are different. For example, for clustering, the pattern space is a probabilistic simplex such that $\mathcal{A} = \{A | A \in \mathbb{R}_+^{n \times k}, A\mathbf{1} = \mathbf{1}\}$ where $\mathbf{1}$ denotes a vector consisting of 1s; i.e., in a clustering pattern matrix A, A_{ij} denotes the probability that the ith object is associated with the jth cluster; the clustering pattern matrix provides the cluster structure of the data and the cluster centers can also be obtained based on the clustering pattern matrix and the data matrix. Another example is spectral dimensionality reduction, in which $\mathcal{A}$ is a subset of an eigen-space such that $\mathcal{A} = \{A \in \mathbb{R}^{n \times k}, A^T A = I\}$ where I denotes an identity matrix, i.e., the pattern matrix A is a low-dimensional embedding of X in the eigen-space.

6.3.1 A General Model for Multiple-View Unsupervised Learning

In this section, we propose a general model for multiple-view unsupervised learning under a distributed framework. Given a multiple-view data set consisting of n objects with m representations denoted as a set of matrices $\mathcal{X} = \{X^{(i)} \in \mathbb{R}^{n \times m_i}\}_{i=1}^m$. Note that $X^{(i)}$ could be either a feature matrix or a graph affinity matrix. When $X^{(i)} \in \mathbb{R}^{n \times m_i}$ is a graph affinity matrix, we have $m_i = n$. Our task is to learn the optimal pattern matrix from $\mathcal{X}$.

First, the appropriate unsupervised learning methods for each representation $X^{(i)}$ are chosen by domain experts to learn a pattern matrix $A^{(i)}$ from $X^{(i)}$. This step is not our focus. In this chapter, we focus on the core problem: learning an optimal pattern $B \in \mathbb{R}^{n \times k}$ from multiple patterns $\mathcal{A} = \{A^{(i)} \in \mathbb{R}^{n \times k_i}\}_{i=1}^m$.

Intuitively, we expect that the optimal pattern is the consensus pattern which is "shared" by multiple patterns in $\mathcal{A}$ as much as possible. From the view of mathematical optimization, we seek the optimal pattern which is close to all the patterns as much as possible under a certain distance measure. Hence, we need to compare two patterns under a certain distance measure. The problem is that since the patterns are learned in an unsupervised way, they are not directly comparable. Let us have an illustrative example under the clustering scenario. Assume that we have two clustering patterns for six data objects, which are represented as $A^{(1)}$ and $A^{(2)}$,

Example 1

$$A^{(1)} = \begin{bmatrix} 1 & 0 & 0 \\ 1 & 0 & 0 \\ 0 & 1 & 0 \\ 0 & 1 & 0 \\ 0 & 0 & 1 \\ 0 & 0 & 1 \end{bmatrix}, A^{(2)} = \begin{bmatrix} 0.5 & 0.5 \\ 0.5 & 0.5 \\ 0.7 & 0.3 \\ 0.6 & 0.4 \\ 0.2 & 0.8 \\ 0.1 & 1.9 \end{bmatrix}.$$

$A^{(1)}$ is a hard clustering pattern with three clusters and $A^{(2)}$ is a soft clustering pattern with two clusters. It is difficult to directly compare $A^{(1)}$ with $A^{(2)}$ to decide how close they are. Note that even for different clustering patterns with the same number of clusters, they are not directly comparable. For example, for the following two clustering patterns,

Example 2

$$A^{(3)} = \begin{bmatrix} 0.8 & 0.2 & 0 \\ 0.8 & 0.2 & 0 \\ 0 & 0.7 & 0.3 \\ 0 & 0.7 & 0.3 \\ 0 & 0.1 & 0.9 \\ 0 & 0.1 & 0.9 \end{bmatrix}, A^{(4)} = \begin{bmatrix} 0.2 & 0 & 0.8 \\ 0.2 & 0 & 0.8 \\ 0.7 & 0.3 & 0 \\ 0.7 & 0.3 & 0 \\ 0.1 & 0.9 & 0 \\ 0.1 & 0.9 & 0 \end{bmatrix},$$

they are both a clustering pattern with three clusters. If we directly compare $A^{(3)}$ with $A^{(4)}$, they appear different. However, the two clusterings are actually equivalent; since cluster 1, 2, and 3 of $A^{(3)}$ actually is equivalent to cluster 3, 1, and 2 of $A^{(4)}$, respectively. This example implies that even two pattern matrices with the same dimension are not directly comparable. This is true for other types of unsupervised learning such as dimensionality reduction.

To solve this problem, we propose to use a mapping function to map a pattern matrix into the pattern space of another pattern matrix; i.e., instead of comparing the two pattern matrices directly, we compare the mapped pattern matrix with another pattern matrix in the same pattern space. For example, for two pattern matrices $A^{(1)} \in \mathbb{R}^{n \times k_1}$ and $A^{(2)} \in \mathbb{R}^{n \times k_2}$, $f(A^{(1)})$ and $A^{(2)}$ are comparable where $f : \mathbb{R}^{n \times k_1} \to \mathbb{R}^{n \times k_2}$ is a mapping function.

By using the concept of the mapping function, we are able to obtain an optimal pattern from multiple patterns existing in different pattern spaces. The basic idea is that the optimal pattern should be "close" to each pattern as much as possible after it is mapped into the pattern space for each pattern. Formally, we define the model of multiple-view unsupervised learning as follows.

DEFINITION 6.1 *Given a set of pattern matrices $\mathcal{A} = \{A^{(i)} \in \mathbb{R}^{n \times k_i}\}_{i=1}^m$, a positive integer k, a set of nonnegative weights $\{w_i \in R_+\}_{i=1}^m$, and a distance function l, the optimal pattern matrix $B \in \mathbb{R}^{n \times k}$ and the* *optimal mapping functions* $\mathcal{F} = \{f_i : \mathbb{R}^{n \times k} \rightarrow \mathbb{R}^{n \times k_i}\}_{i=1}^m$ *are given by the minimization,*

$$\min_{B, \mathcal{F}} \sum_{i=1}^m w_i l(A^{(i)}, f_i(B)). \tag{6.1}$$

In Definition 6.1, we formulate the problem of multiple-view unsupervised learning as an optimization problem w.r.t. both the optimal pattern matrix and the mapping functions. The mapping functions provide not only a technical convenience, but also desirable information in some applications, since they actually denote the relations between the final optimal pattern and the patterns from different views.

A natural question about the model in Definition 6.1 is why we do not use mapping functions to map $A^{(i)}$ into the space of B to have the following model:

$$\min_{B, \mathcal{F}} \sum_{i=1}^m w_i l(f_i(A^{(i)}), B), \tag{6.2}$$

where $\mathcal{F} = \{f_i : \mathbb{R}^{n \times k_i} \rightarrow \mathbb{R}^{n \times k}\}_{i=1}^m$. The above model looks similar to that in Definition Equation 6.1. However, it has a serious problem that there always exists a useless global optimal solution to Equation 6.2, $B^* = \{0\}^{n \times k_i}$ and $\mathcal{F}^* = \{f_i : \mathbb{R}^{n \times k_i} \rightarrow \{0\}^{n \times k}\}_{i=1}^m$; i.e., if we let B be a zero matrix and f_i be a mapping function which maps $A^{(i)}$ to a zero matrix, we always have a minimum distance. Obviously, this solution is useless in real applications. Hence, to design an algorithm to solve the model in Equation 6.2, we need extra constraints to ensure the algorithm not to converge to the region near the useless solution. On the other hand, we do not have this problem with the model in Definition 6.1.

6.3.2 Two Specific Models: Multiple-View Clustering and Multiple-View Spectral Embedding

In this section, we show how to apply the general model of multiple-view learning in Definition 6.1 to two of the most important unsupervised learning settings: clustering and spectral dimensionality reduction.

Definition 6.1 provides a general model which is applicable to different types of unsupervised learning on various multiple-view data. To derive a specific model for a certain type of unsupervised learning, first we need to select the function space for the mapping functions. In this chapter, we adopt the popular linear transformation function as the mapping function, since it provides a computational convenience and has an intuitive meaning in many applications. With mapping functions in the linear function space, i.e., $f_i \in \{f | f(X) = BX\}$, the model is reduced to the following optimization:

$$\min_{B \in \mathbb{R}^{n \times k}, \{P^{(i)} \in \mathbb{R}^{k \times k_i}\}_{i=1}^m} \sum_{i=1}^m w_i l(A^{(i)}, BP^{(i)}). \tag{6.3}$$

In the rest of the chapter, we call $P^{(i)}$ mapping matrix .

Next, based on the properties of pattern spaces in different unsupervised learning tasks, we select appropriate distance functions and constraints on the optimal pattern matrix and mapping matrices.

For multiple-view clustering, each clustering pattern matrix lies in a probability simplex such that $A^{(i)} \in R_+^{n \times k_i}$ and $A\mathbf{1} = \mathbf{1}$. Hence, the optimal clustering pattern matrix B is also in a probabilistic simplex such that $B \in R_+^{n \times k}$ and $B\mathbf{1} = \mathbf{1}$. Since both $A^{(i)}$ and B are nonnegative, we restrict the mapping matrix $P^{(i)}$ to be nonnegative such that $P^{(i)} \in R_+^{k \times k_i}$. This restriction also gives the mapping matrix $P^{(i)}$ an intuitive interpretation, i.e., $P_{pq}^{(i)}$ denotes the weight of the pth cluster of the optimal clustering related to the qth cluster of the ith clustering. For example, the optimal mapping matrix for the two clustering patterns in Example 2 of Section 6.3.1 is

$$P = \begin{bmatrix} 0 & 1 & 0 \\ 0 & 0 & 1 \\ 1 & 0 & 0 \end{bmatrix}.$$

We observe that the weights in P provides the equivalence of the clusters between the two clustering patterns.

For the distance function, since the elements of the pattern matrices denote probabilities, an intuitive choice is KL-divergence or generalized I-divergence. Since generalized I-divergence is more general than KL-divergence in the sense that it does not require that the sum of the elements in a matrix equal to 1, we elect to use generalized I-divergence. Formally, we define the problem of multiple-view clustering as follows.

DEFINITION 6.2 *Given a set of clustering membership matrices* $\mathcal{A} = \{A^{(i)} \in \mathbb{R}_+^{n \times k_i}\}_{i=1}^m$, *a positive integer* k, *and a set of nonnegative weights* $\{w_i \in \mathbb{R}_+\}_{i=1}^m$, *the optimal clustering membership matrix* $B \in \mathbb{R}_+^{n \times k}$ *and the optimal mapping matrices* $\mathcal{P} = \{P^{(i)} \in \mathbb{R}_+^{k \times k_i}\}_{i=1}^m$ *are given by the minimization,*

$$\min_{\substack{B, \mathcal{P} \text{ s.t.} \\ B\mathbf{1}=\mathbf{1}}} \sum_{i=1}^m w_i GI(A^{(i)} \| BP^{(i)}), \tag{6.4}$$

where GI *is generalized I-divergence function such that* $GI(X\|Y) = \sum_{ij}(X_{ij} \log \frac{X_{ij}}{Y_{ij}} - X_{ij} + Y_{ij})$.

One of the most popular dimensionality reduction approach is to embed data objects from high-dimension feature space or a large graph into its low-dimension eigen-space. We call it spectral embedding in this chapter. In multiple view spectral embedding, pattern matrices from multiple-views are spectral embedding matrices such that $A^{(i)} \in \mathbb{R}^{n \times k_i}$ and $A^T A = I$, where I is an identity matrix. Hence, the optimal pattern matrix is also a spectral embedding matrix such that $B \in \mathbb{R}^{n \times k}$ and $B^T B = I$. There are no constraints on the mapping matrices since they serve the purpose of general linear mapping. For the distance function, we elect to use the most popular one, Euclidean distance function. Formally, we define the problem of multiple-view spectral embedding as follows.

DEFINITION 6.3 *Given a set of spectral embedding matrices $\mathcal{A} = \{A^{(i)} \in \mathbb{R}^{n \times k_i}\}_{i=1}^m$, a positive integer k, and a set of nonnegative weights $\{w_i \in R_+\}_{i=1}^m$, the optimal spectral embedding matrix $B \in \mathbb{R}^{n \times k}$ and the optimal mapping matrices $\mathcal{P} = \{P^{(i)} \in \mathbb{R}^{k \times k_i}\}_{i=1}^m$ are given by the minimization,*

$$\min_{\substack{B,\mathcal{P}\text{s.t.} \\ B^T B = I}} \sum_{i=1}^m w_i \|A^{(i)} - BP^{(i)}\|^2, \qquad (6.5)$$

where $\| \cdot \|$ denotes Frobenius norm such that $\|X\|^2 = \sum_{ij} X_{ij}^2$.

Chapter 7

Evolutionary Data Clustering

In may applications, data evolve with time. This is especially typical for relational data. Evolutionary clustering on relational data is extremely challenging, since it involves a complicated system of multiple types of data objects and relations changing over time. In this book, we start with general evolutionary clustering as an initial efforts for evolutionary clustering on relational data. In evolutionary clustering, we need to address two issues at each time step: the current clustering pattern should depend mainly on the current data features; on the other hand, the current clustering pattern should not deviate dramatically from the most recent history, i.e., we expect a certain level of temporal smoothness between clusters in successive time steps. In this chapter, we propose three evolutionary models based on the recent literature on Hierarchical Dirichlet Process (HDP) and Hidden Markov Model (HMM). Those models substantially advance the literature on evolutionary clustering in the sense that not only they both perform better than the existing literature, but more importantly they are capable of automatically learning the cluster numbers and structures during the evolution (for more detailed description of the models refer to [131, 132]).

7.1   Introduction

Evolutionary clustering is a relatively new research topic in data mining. Evolutionary clustering refers to the scenario where a collection of data evolves over the time; at each time, the collection of the data has a number of clusters; when the collection of the data evolves from one time to another, new data items may join the collection and existing data items may disappear; similarly, new clusters may appear and at the same time existing clusters may disappear. Consequently, both the data items and the clusters of the collection may change over the time, which poses a great challenge to the problem of evolutionary clustering in comparison with the traditional clustering. On the other hand, solutions to the evolutionary clustering problem have found a wide spectrum of applications for trend development analysis, social network evolution analysis, and dynamic community development analysis. Potential and existing applications include daily news analysis to observe news focus change, blog analysis to observe community development, and scientific publications analysis to identify the new and hot research directions in a specific area. Due to these important applications, evolutionary clustering has recently become a very hot and focused research topic.

Statistically, each cluster is associated with a certain distribution at each time. A solution to the evolutionary clustering problem is to make an inference to a sequence of distributions from the data at different times.

A reasonable solution to the evolutionary clustering problem must have a clustering result consistent with the original data distribution. Consequently, the following two properties must be satisfied to reflect a reasonable evolutionary clustering problem: (1) The number of clusters as well as the clustering structures at different evolutionary times may change. (2) The clusters of the data between neighboring times should stay the same or have a smooth change, but after a long time, clusters may drift substantially.

In this chapter, we propose a statistical approach to solving the evolutionary clustering problem. We assume that the cluster structure at each time follows a mixture model of the clusters for the data collection at this time; clusters at different times may share common clusters; further, these clusters evolve over the time and some may become more popular while others may become outdated, making the cluster structures and the number of clusters change over the time. Consequently, we use Dirichlet Process (DP) [53] to model the evolutionary change of the clusters over the time. Specifically, we propose three Dirichlet process-based models for the evolutionary clustering problem: DPChain, HDP-EVO, HDP-HTM.

DPChain is based on the Dirichlet Process Mixture (DPM) model [7, 52], which automatically learns the number of the clusters from the evolutionary data; in addition, the cluster mixture proportion information at different times is used to reflect a smooth cluster change over the time. HDP-EVO is developed based on the Hierarchical Dirichlet Process (HDP) model [121] with a set of common clusters on the top level of the hierarchy to explicitly address the cluster correspondence issue in order to solve the evolutionary clustering problem; the middle level is for the clusters at each different time, which are considered as the subsets of the top level clusters; the relationship between the top level clusters and the middle level clusters is obtained through the statistical inference under this model, resulting in explicitly addressing the cluster correspondence issue for the clusters at different times. In the HDP-HTM model, we assume that the cluster structure at each time is a mixture model of the clusters for the data collection at that time; in addition, clusters at different times may share common clusters, resulting in explicitly addressing the cluster-cluster correspondence issue. We adopt the Hierarchical Dirichlet Processes (HDP) [121] with a set of common clusters at the top level of the hierarchy and the local clusters at the lower level at each different times sharing the top level clusters. Further, data and clusters evolve over the time with new clusters and new data items possibly joining the collection and with existing clusters and data items possibly leaving the collection at different times, leading to the cluster structure and the number of clusters evolving over the time. Here, we use the state transition matrix to explicitly reflect the cluster-to-cluster transitions between different times, resulting in explicitly effective solution to the cluster transition correspondence issue. Consequently, we propose the Infinite Hierarchical Hidden Markov State (iH^{2}MS) model to construct the Hierarchical Transition Matrix (HTM) at different times to capture the cluster-to-cluster transition evolution.

7.2 Related Work

Evolutionary Clustering is a recently emerging research topic in data mining. Due to its very short history, there is not much literature on this topic at this time.

[27] were probably considered as the first to address the evolutionary clustering problem in the data mining literature. In their work, a general framework was proposed and two specific clustering algorithms within this framework were developed: evolutionary k-means and evolutionary agglomerative hierarchical clustering. The framework attempted to combine the two properties of evolutionary clustering for the development of these two algorithms: one is the snapshot quality, which measures how well the current data fit the current clustering, and other is the history quality, which measures how smooth the current clustering is with the previous clustering.

Recently, [30] presented an evolutionary spectral clustering approach by incorporating the temporal smoothness constraint into the solution. In order to fit the current data well into the clustering but at the same time not to deviate the clustering from the history too dramatically, the temporal smoothness constraint is incorporated into the overall measure of the clustering quality. Based on the spectral clustering approach, two specific algorithms, PCM and PCQ, were proposed.

These two algorithms were developed by explicitly incorporating the history clustering information into the existing classic clustering algorithm, specifically, k-means, agglomerative hierarchical clustering, and spectral clustering approaches [97,114]. While incorporating the history information into the evolutionary clustering certainly advances the literature on this topic, there is a very restrictive assumption in their work—it is assumed that the number of the clusters over the time stays the same. It is clear that in many applications of evolutionary clustering, this assumption is obviously violated.

Dirichlet process [53] is a statistical model developed in the statistics literature to capture the distribution uncertainties in the space of probability measure. When a random measure is no longer a single distribution, but a mixture distribution, DPM [7,52] is used to extend DP. Statistically, the clustering problem indeed fits into a mixture model, making it natural to use DPM model.

More importantly, DPM allows an infinite number of mixture components, shedding the light on solving the clustering model selection problem. [111] gives a constructive definition of Dirichlet distribution for an arbitrarily measurable base space. This stick-breaking construction is very useful to model the weight of mixture components in the clustering mixture model. Besides the capability of learning the number of clusters from the data automatically, HDP model [121] is further developed for sharing the mixture components across different data collections, making it possible to capture the relationship between the clusters at different times.

Recently in the machine learning community, DP-related models are developed and used to solve the clustering problems such as document topic analysis [21, 121], image clustering [98], and video surveillance activity analysis [126]. [21] developed the Latent Dirichlet Allocation (LDA) model that automatically learns the clustering of topics given a document corpus. However, LDA assumes that the number of clusters is given in advance and is a parametric constant.

[22] designed a family of time series probabilistic models to analyze the evolving topics at different times; they assumed a fixed number of topics and did not consider the clusters' birth or death during the evolution. [62] studied the PANS proceedings by LDA model to identify "hot topics" and "cold topics" by examining temporal dynamics of the documents; they used Bayesian model selection to estimate the number of the topics. [127] presented an LDA-style topic model in which time is an observed continuous variable instead of a Markov discretization assumption. This model is able to capture the trends of the temporal topic evolution; however, the number of topics is still assumed fixed. [140] further developed a time-sensitive Dirichlet process mixture model for clustering documents, which models the temporal correlations between instances. Nevertheless, a strong assumption was made that there is only one cluster and one document at each time, which is too restrictive to handle problems with a collection of clusters and documents at a time. More recently, [130] proposed a statistical model HDP-HMM to provide a solution to evolutionary clusterng, which is able to learn the number of clusters and cluster structure transitions during the evolution.

7.3   Dirichlet Process Mixture Chain (DPChain)

In the following text, boldface symbols are used to denote vectors or matrices, and non-boldface symbols are used to denote scalar variables. Also for

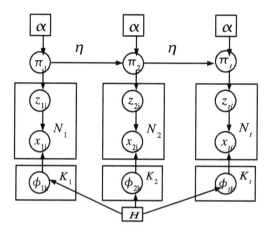

FIGURE 7.1: The DPChain model.

all the variables we have defined, adding a symbol $-s$ either in the subscript or in the superscript to a defined variable means the whole scope the variable is defined for except for the item indicated as s.

The first model we propose is based on the DPM model [7, 52], which is called DPChain model in this paper. For DPChain model, we assume that at each time t a collection of data has K_t clusters and each cluster is derived from a unique distribution. K_t is unknown and is learned from the data. We denote N_t as the number of the data items in this collection at time t.

7.3.1 DPChain Representation

Figure 7.1 illustrates the DPChain model. We use the indicator variable to represent the DPChain model. First we introduce the notations. α denotes the concentration parameter for a Dirichlet distribution. H denotes the base measure of a Dirichlet distribution with the pdf as h. F denotes the distribution of the data with the pdf as f. $\phi_{t,k}$ denotes the parameter of cluster k of the data at time t. At time t, $\phi_{t,k}$ is a sample from distribution H, represented as a parameter of F.

$$\phi_{t,k}|H \sim H$$

$\boldsymbol{\pi}_t$ is the cluster mixutre proportion vector at time t. $\pi_{t,k}$ is the weight of the corresponding cluster k at time t. Consequently, $\boldsymbol{\pi}_t$ is distributed as $stick(\alpha)$ [111] which is described as follows:

$$\boldsymbol{\pi}_t = (\pi_{t,k})_{k=1}^{\infty} \quad \pi_{t,k} = \pi_{t,k}' \prod_{l=1}^{k-1}(1 - \pi_{t,l}') \quad \pi_{t,k}' \sim Beta(1, \alpha) \qquad (7.1)$$

Let $z_{t,i}$ be the cluster indicator at time t for data item i. $z_{t,i}$ follows a multinomial distribution with parameter $\boldsymbol{\pi}_t$.

$$z_{t,i}|\boldsymbol{\pi}_t \sim Mult(\boldsymbol{\pi}_t)$$

Let $x_{t,i}$ denote data item i from the collection at time t. $x_{t,i}$ is modeled as being generated from F with parameter $\phi_{t,k}$ by the assignment $z_{t,i}$.

$$x_{t,i} \mid z_{t,i}, (\phi_{t,k})_{k=1}^{\infty} \sim f(x|\phi_{t,z_{t,i}})$$

In evolutionary clustering, cluster k is smoothly changed from time $t-1$ to t. With this change of the clustering, the number of the data items in each cluster may also change. Consequently, the cluster mixture proportion is an indicator for the population of a cluster. In the classic DPM model, $\boldsymbol{\pi}_t$ represents the cluster mixture. We extend the classic DPM model to the DPChain model by incorporating the temporal information into $\boldsymbol{\pi}_t$. With a cluster smooth change, more recent history has more influence on the current clustering than less recent history. Thus, a cluster with a higher mixture proportion at the present time is more likely to have a higher proportion at the next time. Hence, the cluster mixture at time t may be constructed as follows:

$$\boldsymbol{\pi}_t = \sum_{\tau=1}^{t} \exp\{-\eta(t-\tau)\}\boldsymbol{\pi}_\tau \tag{7.2}$$

where η is a smooth parameter.

This relationship is further illustrated by an extended Chinese Restaurant Process (CRP) [6,20]. We denote $n_{t,k}$ as the number of data items in cluster k at time t, and $n_{t,k}^{-i}$ as the number of data items belonging to cluster k except $x_{t,i}$; $w_{t,k}$ is the smooth prior weight for cluster k at the beginning of time t. According to Equation 7.2, $w_{t,k}$ has the relationship to $n_{\tau,k}$ at the previous time τ:

$$w_{t,k} = \sum_{\tau=1}^{t-1} \exp\{-\eta(t-\tau)\}n_{\tau,k}. \tag{7.3}$$

Then, similar to CRP, the prior probability to sample a data item from cluster k given history assignment $\{\mathbf{z}_1 \ldots \mathbf{z}_{t-1}\}$ and the other assignment at time t, $\mathbf{z}_{t,-i} = \mathbf{z}_t \setminus z_{t,i}$ is as follows:

$$p(z_{t,i} = k|\mathbf{z}_1, ...\mathbf{z}_{t-1}, \mathbf{z}_{t,-i}) \propto$$
$$\begin{cases} \dfrac{w_{t,k}+n_{t,k}^{-i}}{\alpha+\sum_{j=1}^{K_t} w_{t,j}+n_t-1} & \text{if } k \text{ is an existing cluster} \\[3mm] \dfrac{\alpha}{\alpha+\sum_{j=1}^{K_t} w_{t,j}+n_t-1} & \text{if } k \text{ is a new cluster,} \end{cases} \tag{7.4}$$

where $n_t - 1$ is the number of the data items at time t except for $x_{t,i}$, and $x_{t,i}$ is considered as the last data item in the collection at time t. With Equation 7.4 an existing cluster appears again with a probability proportional to $w_{t,k}+n_{t,k}^{-i}$, while a new cluster appears at the first time with a probability proportional to α. If at time t as well as the times before t, the data of cluster k appear infrequently, cluster k has a relatively small weight to appear again in the next time, which leads to a higher probability of becoming death for cluster k. Consequently, this model has the capability to describe the birth or death of a cluster over the evolution. The data item generation process for DPChain model is listed as follows:

1. Sample cluster parameter $\phi_{t,k}$ from the base measure H at each time. The number of the cluster is not a fixed prior parameter but is decided by the data when a new cluster is needed.

2. First, sample the cluster mixture vector $\boldsymbol{\pi}_t$ from $stick(\alpha)$ at each time; then, $\boldsymbol{\pi}_t$ is further smoothly weighted from the exponential sum according to Equation 7.2.

3. At time t, sample the cluster assignment $z_{t,i}$ for data item $x_{t,i}$ from the multinomial distribution with parameter $\boldsymbol{\pi}_t$.

4. Finally, a data item $x_{t,i}$ is generated from distribution $f(x|\phi_{t,z_{t,i}})$ given cluster index variable $z_{t,i}$ and cluster parameter $\phi_{t,k}$.

At each time t, the concentration parameter α may be different. In the sampling process, we just sample α from a Gamma Distribution at each iteration. For a more sophisticated model, α may be modeled as a random variable varying with time, as the rate of generating a new cluster may change over the time.

7.4 HDP Evolutionary Clustering Model (HDP-EVO)

While DPChain model advances the existing literature on evolutionary clustering in the sense that it is capable of learning the cluster numbers over the time, this model fails to have an explicit representation on the cluster correspondence over the time. In order to explicitly capture the cluster correspondence between the data collections of different times, we further develop the HDP Evolutionary Clustering model, which we call HDP-EVO.

7.4.1 HDP-EVO Representation

HDP-EVO model is illustrated in Figure 7.2. Again, we use the indicator variable representation to describe the HDP-EVO model. First, we introduce the notations. γ is the concentration parameter of the Dirichlet distribution of $\boldsymbol{\pi}$. Common clusters for all the collections at different times are shared with

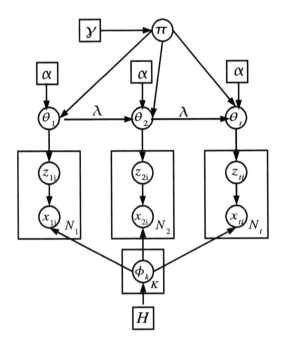

FIGURE 7.2: The HDP-EVO model.

the global cluster set with mixture proportion vector $\boldsymbol{\pi}$.

$$\boldsymbol{\pi}|\gamma \sim stick(\gamma)$$

ϕ_k is the parameter for a cluster with i.i.d. sampled from a distribution H.

$$\phi_k|H \sim H.$$

The clusters appearing at time t are a subset of the common clusters with a local cluster mixture parameter vector $\boldsymbol{\theta}_t$.

$$\boldsymbol{\theta}_t|\alpha, \boldsymbol{\pi} \sim DP(\alpha, \boldsymbol{\pi}),$$

where α is the concentration parameter. At different times, a different $\boldsymbol{\theta}_t$ shares the common global clusters which establish the correspondence between the local clusters at different times and the global clusters.

Similar to DPChain model, the mixture proportion of the clusters evolves over the time, favoring recent history. We assume again an exponential smooth transition:

$$\boldsymbol{\theta}_t = \sum_{\tau=1}^{t} \exp\{-\lambda(t-\tau)\}\boldsymbol{\theta}_\tau, \tag{7.5}$$

where λ is a smooth parameter. We denote $z_{t,i}$ as the cluster assignment at time t for the data item $x_{t,i}$, and follow a multinomial distribution of $\boldsymbol{\theta}_t$.

$$z_{t,i}|\boldsymbol{\theta}_t \sim Mult(\boldsymbol{\theta}_t)$$

Finally, $x_{t,i}$ is modeled as being drawn from the distribution F with the parameter ϕ_k under cluster k.

$$x_{t,i} \mid z_{t,i}, (\phi_k)_{k=1}^{\infty} \sim f(x|\phi_{z_{t,i}}).$$

Now, the data generation process is described as follows:

1. The common global clusters' parameter vector $\boldsymbol{\phi}$ is sampled from distribution H. The number of the cluster is not a fixed prior but is decided by the data when a new cluster is needed.

2. Sample global cluster mixture proportion $\boldsymbol{\pi}$ from $stick(\gamma)$.

3. At time t, first sample the local clusters' mixture proportion vector $\boldsymbol{\theta}_t$ from $DP(\alpha, \boldsymbol{\pi})$; then do smoothly weighted sum according to Equation 7.5.

4. $z_{t,i}$, the assignment of the cluster for $x_{t,i}$, is sampled from the multinomial distribution with parameter $\boldsymbol{\theta}_t$.

5. Finally, we sample $x_{t,i}$ from distribution F with parameter ϕ_k, given the cluster assignment $z_{t,i} = k$.

Based on the above generation process, the cluster number can be automatically learned through the inference from the data at each time. All the local clusters at different times are capable of establishing a correspondence relationship among themselves from the top level of the commonly shared global clusters. With the introduction of the exponentially weighted smoothness of the mixture proportion vector at different times, the cluster may smoothly evolve over the time.

7.4.2 Two-Level CRP for HDP-EVO

The indicator variable representation of HDP-EVO directly assigns clusters to data. In order to design the Gibbs sampling process for HDP-EVO, we further illustrate HDP-EVO model as a two-level CRP.

Under the standard CRP model [6,20], each table corresponds to one cluster. Here, we further categorize the clusters into a higher level, global clusters that are commonly shared across all data collections at different times, and the lower level, local clusters, i.e., the tables of a Chinese Restaurant with data items sitting around, at each time. We use k to denote the k-th global cluster and use tab to denote the tab-th local cluster.

At each time t, the data collection is modeled as being generated from the local clusters with the parameters $\{\psi_{t,1}, \ldots, \psi_{t,tab}, \ldots\}$, each of which is sampled from the commonly shared global clusters with parameters $\{\phi_1, \ldots, \phi_k, \ldots\}$ in the CRP style [6, 20]. We use $tab_{t,i}$ to denote the table (i.e., the local cluster) at time t for $x_{t,i}$. We assign global cluster k to table tab, if all the data clustered into local cluster tab at time t are distributed with parameter ϕ_k. We explicitly introduce $k_{t,tab}$ to represent this mapping relationship. Similarly, we introduce $tab_{t,i}$ to denote the mapping that $x_{t,i}$ is clustered into table tab at time t. Let $n_{t,tab}$ be the number of the data items at table tab at time t, $n_{t,tab}^{-i}$ be the number of the data items in table tab except for $x_{t,i}$, and n_t be the total number of the data items at time t. Let $m_{t,k}$ be the number of the tables at time t belonging to the global cluster k, $m_{t,k}^{-tab}$ be number of the tables in cluster k except for tab, and m_t be the total number of the tables at time t,

Under the two-level CRP, at time t, we first sample which table tab $x_{t,i}$ belongs to, given the history $\{tab_{t,1}, \ldots, tab_{t,i-1}\}$ in which by the exchangeability $x_{t,i}$ may be considered as the last data item at time t:

$$p(tab_{t,i}|tab_{t,1}, \ldots, tab_{t,i-1}) \propto$$
$$\begin{cases} \frac{n_{t,tab}^{-i}}{\alpha + n_t - 1} & \text{if } tab \text{ is an existing table} \\ \frac{\alpha}{\alpha + n_t - 1} & \text{if } tab \text{ is a new table,} \end{cases} \tag{7.6}$$

where α is the concentration parameter. To ensure the smooth transition over the history, we also denote $w_{t,k}$ as the smooth prior weight for cluster k at time t. Thus, we have

$$w_{t,k} = \sum_{\tau=1}^{t-1} \exp\{-\lambda(t - \tau)\} m_{t,k}, \tag{7.7}$$

Denoting $\mathbf{K}$ as the all the history global cluster assignment mapping up to time t inclusive, the likelihood of having the assignment mapping $k_{t,tab}$ is

$$p(k_{t,tab}|\mathbf{K} \setminus k_{t,tab}) \propto$$
$$\begin{cases} \frac{m_{t,k}^{-tab} + w_{t,k}}{\gamma + m_t - 1 + \sum_{j=1}^{K_t} w_{t,j}} & \text{if } k \text{ is an existing cluster} \\ \frac{\gamma}{\gamma + m_t - 1 + \sum_{j=1}^{K_t} w_{t,j}} & \text{if } k \text{ is a new cluster,} \end{cases} \tag{7.8}$$

where γ is the concentration parameter.

7.5   Infinite Hierarchical Hidden Markov State Model

Here, we propose a new Infinite Hierarchical Hidden Markov State Model for Hierarchical Transition Matrix (HTM) and provide an update construction

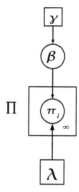

FIGURE 7.3: The iH^{2}MS model.

scheme based on this model. Figure 7.3 illustrates this model.

7.5.1   iH^{2}MS Representation

Tradtionally, Hidden Markov Model (HMM) has a *finite* state space with K hidden states, say $\{1, 2, \ldots K\}$. For the hidden state sequence $\{s_1, s_2, \ldots s_T\}$ up to time T, there is a K by K state transition probability matrix $\mathbf{\Pi}$ governed by Markov dynamics with all the elements $\pi_{i,j}$ of each row $\boldsymbol{\pi}_i$ summed to 1.

$$\pi_{i,j} = p(s_t = j | s_{t-1} = i).$$

The initial state probability for state i is $p(s_1 = i)$ with the summation of all the initial probabilities equal to 1. For observation x_t in the observation sequence $\{x_1, x_2, \ldots x_T\}$, given state $s_t \in \{1, 2, \ldots, K\}$, there is a parameter ϕ_{s_t} drawn from the base measure H which parameterizes the observation likelihood probability.

$$x_t | s_t \sim F(\phi_{s_t}).$$

However, when dealing with a countable *infinite* state space, $\{1, 2, \ldots K, \ldots\}$, we must adopt a new model similar to that in [15] for a state transition probability matrix with an *infinite* matrix dimension. Thus, the dimension of the state transition probability matrix now has become infinite. $\boldsymbol{\pi}_i$, the ith row of the transition probability matrix $\mathbf{\Pi}$, may be represented as the mixing proportions for all the next infinite states, given the current state. Thus, we model it as a DP with an infinite dimension with the summation of all the elements in a row normalized to 1, which leads to an infinite number of DPs' construction for an infinite transition probability matrix.

With no further prior knowledge on the state sequence, a typical prior for the transition probability may be the symmetric Dirichlet distributions. Similar to [121], we intend to construct a hierarchical Dirichlet model to keep different rows of the transition probability matrix to share part of the prior mixing proportions of each state at the top level. Consequently, we adopt a new state model, Infinite Hierarchical Hidden Markov State (iH^{2}MS) model, to construct the Infinite Transition Probability Matrix which is called the Hierarchical Transition Matrix (HTM).

Similar to HDP [121], we draw a random probability measure on the infinite state space $\boldsymbol{\beta}$ as the top level prior from $stick(\gamma)$ represented as the mixing proportions of each state.

$$\boldsymbol{\beta} = (\beta_k)_{k=1}^{\infty} \quad \beta_k = \beta_k' \prod_{l=1}^{k-1}(1 - \beta_l') \quad \beta_k' \sim Beta(1, \gamma). \tag{7.9}$$

Here, the mixing proportion of state k, β_k, may also be interpreted as the prior mean of the transition probabilities leading to state k. Hence, $\boldsymbol{\beta}$ may be represented as the prior random measure of a transition probability DP.

For the ith row of the transition matrix $\boldsymbol{\Pi}$, $\boldsymbol{\pi}_i$, we sample it from $DP(\lambda, \boldsymbol{\beta})$ with a smaller concentration parameter λ implying a larger variability around the mean measure $\boldsymbol{\beta}$. The stick-breaking representation for $\boldsymbol{\pi}_i$ is as follows:

$$\boldsymbol{\pi}_i = (\pi_{i,k})_{k=1}^{\infty} \quad \pi_{i,k} = \pi_{i,k}' \prod_{l=1}^{k-1}(1 - \pi_{i,l}') \quad \pi_{i,k}' \sim Beta(1, \lambda). \tag{7.10}$$

Specifically, $\pi_{i,k}$ is the state transition probability from the previous state i to the current state k as $p(s_t = k|s_{t-1} = i)$.

Now, each row of the transition probability matrix is represented as a DP which shares the same reasonable prior on the mixing proportions of the states. For a new row corresponding to a new state k, we simply draw a transition probability vector $\boldsymbol{\pi}_k$ from $DP(\lambda, \boldsymbol{\beta})$, resulting in constructing a countably infinite transition probability matrix.

7.5.2   Extention of iH^{2}MS

The transition probability constructed by iH^{2}MS may be further extended to the scenario where there are more than one state at each time. Suppose that there is a countably infinite global state space $\boldsymbol{S} = \{1, 2, \ldots, K, \ldots\}$ including states in all the state space $\boldsymbol{S}_t$ at each time t, where $\boldsymbol{S}_t \subseteq \boldsymbol{S}$. For any state $s_t \in \boldsymbol{S}_t$ at time t and state $s_{t-1} \in \boldsymbol{S}_{t-1}$ at time $t-1$, we may adopt the transition probability $\pi_{i,k}$ to represent $p(s_t = k|s_{t-1} = i)$. Similarly, $\pi_{i,k}$ here still has the property that there is a natural tendency for a transition to appear more frequently at the current time if such a transition appears more frequently at a previous time. Therefore, it is reasonable to model a row of transition as a DP with an infinite dimension. We will discuss this extension in detail later.

7.5.3 Maximum Likelihood Estimation of HTM

Let X be the observation sequence, which includes all the observations X_t at each time t, where $X_t \subseteq X$. Now, the question is how to represent the countably infinite state space in a Hierarchical Transition Matrix (HTM). Notice that, at each time, there is in fact a finite number of observations X_t; the state space S_t at each time t must be arbitrarily finite even though conceptually the global state space S may be considered as countably infinite. Further, we adopt the stick-breaking representation for the DP prior [70,121] to iteratively handle an arbitrary number of the states and accordingly the transition probability matrix up to time t.

Suppose that up to time t there are K current states and we use $K+1$ to index a potentially new state. Then $\boldsymbol{\beta}$ may be represented as:

$$\boldsymbol{\beta} = \{\beta_1, \ldots \beta_K, \beta_u\} \quad \beta_u = \sum_{k=K+1}^{\infty} \beta_k \quad \sum_{k=1}^{K} \beta_k + \beta_u = 1. \tag{7.11}$$

Given $\boldsymbol{\beta}$, the Dirichlet prior measure of ith row of the transition probability matrix $\boldsymbol{\pi}_i$ has a dimension $K+1$. The last element β_u is the prior measure of the transition probability from state i to an unpresented state u. The prior distribution of $\boldsymbol{\pi}_i$ is $Dir(\lambda\beta_1, \ldots \lambda\beta_K, \lambda\beta_u)$.

When a new state is instantiated, we sample b from $Beta(1, \gamma)$, and set the new proportions for the new state $K^{new} = K+1$ and another potentially new state $K^{new} + 1$ as:

$$\beta_{K^{new}} = b\beta_u \quad \beta_u^{new} = (1-b)\beta_u. \tag{7.12}$$

Now, K is updated as K^{new}, β_u as β_u^{new}, and the number of the states may continue to increase if yet another new state is instantiated, resulting in a countably infinite transition probability matrix.

Since up to time t, there are K states in the hidden state space, it is possible to adopt Baum-Welch algorithm [90, 101] to estimate the posterior of the transition probability matrix $\boldsymbol{\Pi}$ by the maximum likelihood Optimization. Similar to [90], we have a Dirichlet prior on each row of $\boldsymbol{\Pi}$, $\boldsymbol{\pi}_i$. Consequently, we have the M-step for updating π_{ik}, according to the standard Baum-Welch optimization equation:

$$\pi_{i,k} = p(s_t = k | s_{t-1} = i) = \begin{cases} \frac{n_{i,k}^t + \lambda\beta_k}{n_i^t + \lambda} & k \text{ is an exisiting state} \\ \frac{\lambda\beta_u}{n_i^t + \lambda} & k \text{ is a new state,} \end{cases} \tag{7.13}$$

where approximately $n_{i,k}^t$ is the expected number of the transitions from state i to state k up to time t, n_i^t is the expected number of the transitions out of state i up to time t. For a state of any common observation x between two adjacent times τ and $\tau+1$ up to time t,

$$n_{i,k}^t = \sum_{\tau=1}^{t-1} \delta(s_{\tau,x}, i)\delta(s_{\tau+1,x}, k) \quad n_i^t = \sum_{k=1}^{K} n_{i,k}^t$$

where $s_{\tau,x}$ and $s_{\tau+1,x}$ capture the correpondence between the states with the same data item at adjacent times. Here we use Kronecker-delta function ($\delta(a,b) = 1$ iff $a = b$ and 0 otherwise) to count the number of the state transitions for all the common observations up to time t.

Conceptually, in Equation 7.13 we may consider $\lambda\beta_k$ as the pseudo-observation of the transition from state i to k (i.e., the strength of the belief for the prior state transition), and $\lambda\beta_u$ as the probability of a new state transitioned from state i. Equation 7.13 is equivalent to Blackwell-MacQueen urn Scheme [20] to sample a state. Thus, this posterior maximum likelihood estimation of the transition probability to a state is equal to probability of such a state posterior sampled by the polya urn scheme [20] given a sequence of the states and observations up to time t.

7.6   HDP Incorporated with HTM (HDP-HTM)

To capture the state(cluster) transition correspondence during the evolution at different times, we propose the HTM; at the same time, we must capture the state-state (cluster-cluster) correspondence, which may be handled by a hierarchical model with the top level corresponding to the global states [1] and the lower level corresponding to the local states, where it is natural to model the statistical process as HDP [121]. Consequently, we propose to combine HDP with HTM as a new HDP-HTM model, as illustrated in Figure 7.4.

7.6.1   Model Representation

Let the global state space S denote the global cluster set, which includes all the states $S_t \subseteq S$ at all the times t. The global observation set X includes all the observations X_t at each time t, of which each data item i is denoted as $x_{t,i}$.

We draw the global mixing proportion from the global states β with the stick-breaking representation using the concentration parameter γ from Equation 7.9. The global measure G_0 may be represented as:

$$G_0 = \sum_{k=1}^{\infty} \beta_k \delta_{\phi_k},$$

where ϕ_k is drawn from the base probability measure H with pdf h, and δ_{ϕ_k} is the concentration measure on ϕ_k.

Different from HDP, here we must consider the evolution of the data and the states (i.e., the clusters). The distribution of the clusters at time t is not

---

[1]Each state is represented as a distinct cluster.

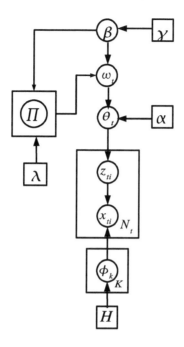

FIGURE 7.4: The HDP-HTM model.

only governed by the global measure G_0, but also is controlled by the data and cluster evolution in the history. Consequently, we make an assumption that the data and the clusters at time t are generated from the previous data and clusters, according to the mixture proportions of each cluster and the transition probability matrix. The global prior mixture proportions for the clusters is $\boldsymbol{\beta}$, and the state transition matrix $\boldsymbol{\Pi}$ provides the information of the previous state evolution in the history up to time t. Now, the expected number of the data items generated by cluster k is proportional to the number of data items in the clusters in the history multiplied by the transition probabilities from these clusters to state k; specificially, the mean mixture proportion for cluster k at time t, $\boldsymbol{\omega}_t$, is defined as follows:

$$\omega_{t,k} = \sum_{j=1}^{\infty} \beta_j \pi_{j,k}.$$

More precisely, $\boldsymbol{\omega}_t$ is further obtained by

$$\boldsymbol{\omega}_t = \boldsymbol{\beta} \cdot \boldsymbol{\Pi}. \tag{7.14}$$

Clearly, by the transition probability property, $\sum_{k=1}^{\infty} \omega_{t,k} = 1$, $\sum_{k=1}^{\infty} \pi_{i,k} =$

1, and the stick-breaking property $\sum_{j=1}^{\infty} \beta_j = 1$.

$$\sum_{k=1}^{\infty} \omega_{t,k} = \sum_{k=1}^{\infty}\sum_{j=1}^{\infty} \beta_j \pi_{j,k} = \sum_{j=1}^{\infty} \beta_j \sum_{k=1}^{\infty} \pi_{j,k} = \sum_{j=1}^{\infty} \beta_j = 1.$$

Thus, the mean mixture proportion ω_t may be taken as the new probability measure at time t on the global cluster set. With the concentration parameter α, we draw the mixture proportion vector θ_t from $DP(\alpha, \omega_t)$.

$$\theta_t | \alpha, \omega_t \sim DP(\alpha, \omega_t).$$

Now, at time t, the local measure G_t shares the global clusters parameterized by $\phi = (\phi_k)_{k=1}^{\infty}$ with the mixing proportion vector θ_t.

$$G_t = \sum_{k=1}^{\infty} \theta_{t,k} \delta_{\phi_k}.$$

At time t, given the mixture proportion of the clusters θ_t, we draw a cluster indicator $z_{t,i}$ for data item $x_{t,i}$ from a multinomial distribution:

$$z_{t,i} | \theta_t \sim Mult(\theta_t).$$

Once we have the cluster indicator $z_{t,i}$, data item $x_{t,i}$ may be drawn from distribution F with pdf f, parameterized by ϕ from the base measure H.

$$x_{t,i} | z_{t,i}, \phi \sim f(x | \phi_{z_{t,i}}).$$

Finally, we summarize the data generation process for HDP-HTM as follows:

1. Sample the cluster parameter vector ϕ from the base measure H. The number of the parameters is unknown *a priori*, but is determined by the data when a new cluster is needed.

2. Sample the global cluster mixture vector β from $stick(\gamma)$.

3. At time t, compute the mean measure ω_t for the global cluster set by β and Π according to Equation 7.14.

4. At time t, sample the local mixture proportion θ_t by $DP(\alpha, \omega_t)$.

5. At time t, sample the cluster indicator $z_{t,i}$ from $Mult(\theta_t)$ for data item $x_{t,i}$.

6. At time t, sample data item $x_{t,i}$ from $f(x | \phi_{z_{t,i}})$ given cluster indicator $z_{t,i}$ and parameter vector ϕ.

Part II

Algorithms

Chapter 8

Co-Clustering

In Chapter 2, we presented a new co-clustering framework, Block Value Decomposition (BVD), for bi-type heterogeneous relational data, which factorizes the relational data matrix into three components the row-coefficient matrix R, the block value matrix B, and the column-coefficient matrix C. In this chapter, under BVD framework, we focus on a special yet very popular case—nonnegative relational data, and propose a specific novel co-clustering algorithm that iteratively computes the three decomposition matrices based on the multiplicative updating rules.

8.1 Nonnegative Block Value Decomposition Algorithm

The objective function in Equation 2.2 is convex in R, B, and C, respectively. However, it is not convex in all of them simultaneously. Thus, it is unrealistic to expect an algorithm to find the global minimum. We derive an EM [1] style algorithm that converges to a local minimum by iteratively updating the decomposition using a set of multiplicative updating rules.

First, we prove the following theorem which is the basis of the Nonnegative Block Value Decomposition (NBVD) algorithm.

THEOREM 8.1

If R, B and C are a local minimizer of the objective function in (2.2), then the equations

$$(ZC^T B^T) \odot R - (RBCC^T B^T) \odot R = 0 \tag{8.1}$$

$$(R^T ZC^T) \odot B - (R^T RBCC^T) \odot B = 0 \tag{8.2}$$

$$(B^T R^T Z) \odot C - (B^T R^T RBC) \odot C = 0 \tag{8.3}$$

are satisfied, where $\odot$ denotes the Hadamard product or entrywise product of two matrices.

75

PROOF   The objective function f can be expanded as follows:

$$
\begin{aligned}
f(R, B, C) &= \mathrm{tr}((Z - RBC)(Z - RBC)^T) \\
&= \mathrm{tr}(ZZ^T - 2ZC^T B^T R^T + RBCC^T B^T R^T) \\
&= \mathrm{tr}(ZZ^T) - 2\mathrm{tr}(2ZC^T B^T R^T) \\
&\quad + \mathrm{tr}(RBCC^T B^T R^T)).
\end{aligned}
\tag{8.4}
$$

where the second step of the derivation uses the communicative property of a matrix trace.

Let λ_1, λ_2, and λ_3 be the Lagrange multipliers for the constraint R, B, and $C \geq 0$, respectively, where $\lambda_1 \in \Re^{k \times n}$, $\lambda_2 \in \Re^{l \times k}$, and $\lambda_3 \in \Re^{m \times l}$. The Lagrange function
$L(R, B, C, \lambda_1, \lambda_2, \lambda_3)$ becomes

$$
L = f(R, B, C) - \mathrm{tr}(\lambda_1 R^T) - \mathrm{tr}(\lambda_2 B^T) - \mathrm{tr}(\lambda_3 C^T).
\tag{8.5}
$$

The Kuhn-Tucker (KKT) conditions are

$$
\partial L / \partial R = 0 \tag{8.6}
$$
$$
\partial L / \partial B = 0 \tag{8.7}
$$
$$
\partial L / \partial C = 0 \tag{8.8}
$$
$$
\lambda_1 \odot R = 0 \tag{8.9}
$$
$$
\lambda_2 \odot B = 0 \tag{8.10}
$$
$$
\lambda_3 \odot C = 0 \tag{8.11}
$$

Substituting Equation 8.4 into Equation 8.5 and taking the derivatives, we obtain the following three equations from Equations 8.6 through 8.8, respectively:

$$
2ZC^T B^T - 2RBCC^T B^T + \lambda_1 = 0 \tag{8.12}
$$
$$
2R^T ZC^T - 2R^T RBCC^T + \lambda_2 = 0 \tag{8.13}
$$
$$
2B^T R^T Z - 2B^T R^T RBC + \lambda_3 = 0 \tag{8.14}
$$

Applying the Hadamard multiplication on both sides of Equations 8.12 through 8.14 by R, B, and C, respectively, and using Conditions Equation 8.9 through 8.11, the proof is completed.   ⬚

As a nonlinear optimization problem, NBVD may be solved using the traditional numeric methods such as Newton-type methods and gradient methods. These methods have their pros and cons. Some are simple but slow in convergence, such as the gradient descent. Others are fast in convergence but are complicated, such as the conjugate gradient. Here we propose a set of simple multiplicative updating rules as a solution to NBVD. Compared with the gradient method, these multiplicative updating rules do not require the step-size parameter which can be inconvenient for applications involving large data sets because the convergence is sensitive to the choice of the step-size parameter. Based on Theorem 8.1, we propose following updating rules:

$$R_{ij} \leftarrow R_{ij} \frac{(ZC^T B^T)_{ij}}{(RBCC^T B^T)_{ij}} \tag{8.15}$$

$$B_{ij} \leftarrow B_{ij} \frac{(R^T ZC^T)_{ij}}{(R^T RBCC^T)_{ij}} \tag{8.16}$$

$$C_{ij} \leftarrow C_{ij} \frac{(B^T R^T Z)_{ij}}{(B^T R^T RBC)_{ij}} \tag{8.17}$$

Based on these updating rules, the NBVD algorithm is listed in Algorithm 1. In Step 2 of Algorithm 1, ϵ is a very small positive number used to avoid

---

Algorithm 1 NBVD co-clustering algorithm

1: Initialize R, B, and C with nonnegative values.
2: Update R, B, and C until convergence or over the maximum number of iterations:

$$R_{ij} \leftarrow R_{ij} \frac{(ZC^T B^T)_{ij} + \epsilon}{(RBCC^T B^T)_{ij} + \epsilon}$$

$$B_{ij} \leftarrow B_{ij} \frac{(R^T ZC^T)_{ij} + \epsilon}{(R^T RBCC^T)_{ij} + \epsilon}$$

$$C_{ij} \leftarrow C_{ij} \frac{(B^T R^T Z)_{ij} + \epsilon}{(B^T R^T RBC)_{ij} + \epsilon}$$

3: Normalize the resulting matrices from Step 2 by a predefined normalization procedure.
4: Assign the ith row of Z to cluster $\hat{x}$ where $\hat{x} = \arg\min_j R_{ij}$.
5: Assign the jth column of Z to cluster $\hat{y}$ where $\hat{y} = \arg\min_i C_{ij}$.

---

dividing by 0.

The time complexity of the NBVD algorithm can be shown as $\mathcal{O}(t(k + l)nm)$ where t is the number of the iterations. The complexity is the same as that of the classic one-way clustering algorithm, k-means clustering whose time complexity is $\mathcal{O}(tknm)$. Since NBVD algorithm is simple to implement and only involves basic matrix operations, it is easy to take the benefit of a distributed computing when dealing with a very large data set.

Finally, we consider a special case of NBVD. In practice, there exists a special type of data, symmetric dyadic data. The notion of symmetric dyadic refers to a domain with two identical sets of objects , $\mathcal{X} = \{x_1, \ldots, x_n\}$, in which the observations are made for $dyads(a, b)$, where both a and b are from X and $dyads(a, b) = dyads(b, a)$. Symmetric dyadic data may be considered as a two-dimensional symmetric matrix. For example, a proximity matrix may be considered as a symmetric dyadic data.

8.2   Proof of the Correctness of the NBVD Algorithm

The conditions in Theorem 8.1 are the necessary conditions but not the sufficient conditions for a local minimum. In order to show that the NBVD algorithm is correct, i.e., the NBVD algorithm always converges to a local minimum, we prove the following theorem.

THEOREM 8.2
The objective function Equation 2.2 is nonincreasing under the updating rules in Equations 8.15 through 8.17.

To prove Theorem 8.2, we make use of an auxiliary function similar to that used in the EM algorithm [1, 108] and NMF [83].

DEFINITION 8.1   *$G(c, c^t)$ is an auxiliary function for $F(c)$ if the conditions*

$$G(c, c^t) \geq F(c) \qquad (8.18)$$
$$G(c, c) = F(c) \qquad (8.19)$$

are satisfied. t denotes the discrete time index.

The auxiliary function is an important and useful concept due to the following lemma.

LEMMA 8.1
If G is an auxiliary function, then F is nonincreasing under the updating rule

$$c^{t+1} = \arg\min_c G(c, c^t). \qquad (8.20)$$

PROOF $F(c^{t+1}) \leq G(c^{t+1}, c^t) \leq G(c^t, c^t) = F(c).$ ☐

By defining the appropriate auxiliary functions for the objective function in Equation 2.2, Theorem 8.2 follows from Lemma 8.1. Lemma 8.2 below gives a simple auxiliary function for the updating rule given in Equation 8.17.

Based on the observation that the objective function in Equation 2.2 is separable in the columns of C, we can update each column of C without considering the others. Thus, we simplify the problem to the case of a single column of C, denoted as $c \in \Re^l$. The corresponding column of Z is denoted as $z \in \Re^n$.

LEMMA 8.2
Let

$$\mu = \frac{c \odot c}{c^t} \tag{8.21}$$

where the division between vectors is entrywise division. Then

$$G(c, c^t) = z^T z - 2c^T B^T R^T z + \mu^T B^T R^T RBc^t \tag{8.22}$$

is an auxiliary function for

$$F(c) = \|z - RBc\|^2. \tag{8.23}$$

PROOF Equation 8.23 can be rewritten as

$$\begin{aligned} F(c) &= (z - RBc)^T (z - RBc) \\ &= z^T z - 2c^T B^T R^T z + c^T B^T R^T RBc. \end{aligned} \tag{8.24}$$

When $c = c^t$, we have $\mu = c$. Thus $G(c, c) = F(c)$. To show $G(c, c^t) \geq F(c)$, we compare Equation 8.22 with Equation 8.24 to find that it is equivalent to show

$$\triangle = \mu^T B^T R^T RBc^t - c^T B^T R^T RBc \geq 0. \tag{8.25}$$

For convenience, let $Q = B^T R^T RB$, hence Q is symetric. We prove Equation 8.25 as follows:

$$\begin{aligned} \triangle &= \mu^T Q c^t - c^T Q c \\ &= \sum_{i,j} \mu_i Q_{ij} c_j^t - \sum_{i,j} c_i Q_{ij} c_j \\ &= \sum_{i,j} Q_{ij} (\mu_i c_j^t - c_i c_j) \\ &= \sum_{i,j} Q_{ij} (c_i^2 c_j^t / c_i^t - c_i c_j) \\ &= \sum_{i<j} (Q_{ij} (c_i^2 c_j^t / c_i^t - c_i c_j) + Q_{ji} (c_j^2 c_i^t / c_j^t - c_j c_i)) \\ &= \sum_{i<j} \frac{Q_{ij}}{c_i^t c_j^t} ((c_i c_j^t)^2 + (c_j c_i^t)^2 - 2c_i c_j^t c_j c_i^t) \\ &= \sum_{i<j} \frac{Q_{ij}}{c_i^t c_j^t} (c_i c_j^t - c_j c_i^t)^2 \\ &\geq 0 \end{aligned}$$

where $1 \leq i \leq k$ and $1 \leq j \leq l$.   ☐

The following two lemmas define the auxiliary functions with respect to B and r, respectively, where r^t represents a row of R.

LEMMA 8.3
Let

$$H = \frac{B \odot B}{B^t} \tag{8.26}$$

where the division between two matrices is entrywise division. Then

$$
\begin{aligned}
G(B, B^t) = \operatorname{tr}(Z^T Z) &- 2\operatorname{tr}(R^T Z C^T B^T) \\
&+ \operatorname{tr}(R^T R B^t C C^T H^T)
\end{aligned}
\tag{8.27}
$$

is an auxiliary function for

$$F(B) = \|Z - RBC\|^2. \tag{8.28}$$

LEMMA 8.4
Let

$$\nu = \frac{r \odot r}{r^t} \tag{8.29}$$

where the division between two vectors is entrywise division. Then

$$G(r, r^t) = z^T z - 2 z^T C^T B^T r + r^{t^T} B C R^T B^t \nu \tag{8.30}$$

is an auxiliary function for

$$F(r) = \|z^T - r^T BC\|^2. \tag{8.31}$$

Lemmas 8.3 and 8.4 may be proven similarly to that of Lemma 8.2 and we omit the proofs here.

Now we are ready to prove Theorem 8.2.

PROOF   First, Equation 8.22 can be rewritten as

$$G(c, c_t) = \sum_i \left(z_i^2 - 2(B^T R^T z)_i c_i + \frac{(B^T R^T R B c^t)_i c_i^2}{c_i^t} \right) \tag{8.32}$$

The derivative of $G(c, c_t)$ with respect to c_i is

$$\frac{\partial G}{\partial c_i} = -2(B^T R^T z)_i + 2 \frac{(B^T R^T R B c^t)_i c_i}{c_i^t} \tag{8.33}$$

Then, by Lemma 8.1, $F(c)$ is nonincreasing under the updating rule:

$$c_i^{t+1} = c_i \frac{(B^T R^T z)_i}{(B^T R^T R B c^t)_i} \tag{8.34}$$

Similarly, F can be shown to be nonincreasing under the updating rules for R and B. □

Chapter 9

Heterogeneous Relational Data Clustering

In Chapter 3, we proposed a general model, the relation summary network, to find the hidden structures (the local cluster structures and the global community structures) from a k-partite heterogeneous relation graph. The model provides a principal framework for unsupervised learning on k-partite heterogeneous relation graphs of various structures. In this chapter, we derive a novel algorithm to identify the hidden structures of a k-partite heterogeneous relation graph by constructing a relation summary network to approximate the original k-partite heterogeneous relation graph under a broad range of distortion measures. We also establish the connections between existing clustering approaches and the proposed model to provide a unified view to the clustering approaches.

9.1 Relation Summary Network Algorithm

In this section, we derive an iterative algorithm to find the Relation Summary Network (RSN) (local optima) for a k-partite heterogeneous relation graph. It can be shown that the RSN problem is NP-hard (the proof is omitted here); hence, it is not realistic to expect an efficient algorithm to find the global optima.

First we reformulate the RSN problem based on the matrix representation of a k-partite heterogeneous relation graph. Given a k-partite $G = (V_1, \ldots, V_m, E)$, the weights of edges between V_i and V_j can be represented as a matrix $A^{(ij)} \in \mathbb{R}^{n_i \times n_j}$, where $n_i = |V_i|$, $n_j = |V_j|$, and $A_{hl}^{(ij)}$ denotes the weight of the edge (v_{ih}, v_{jl}), i.e., $e(v_{ih}, v_{jl})$. Similarly in an RSN $G^s = (V_1, \ldots, V_m, S_1, \ldots, S_m, E^s)$, $B^{(ij)} \in \mathbb{R}^{k_i \times k_j}$ denotes the weights of edges between S_i and S_j, i.e., $B_{pq}^{(ij)}$ denotes $e^s(s_{ip}, s_{jq})$; $C^{(i)} \in \{0,1\}^{n_i \times k_i}$ denotes the weights of edges between V_i and S_i, i.e., $C^{(i)}$ is an indicator matrix such that if $e^s(v_{ih}, s_{ip}) = 1$, then $C_{hp}^{(i)} = 1$. Hence, we represent a k-partite as a set of matrices. Note that under the RSN model, we do not use one graph affinity matrix to represent the whole graph as in the graph

TABLE 9.1:   A list of Bregman divergences and the corresponding convex functions

Name	$D_\phi(x, y)$	$\phi(x)$
Euclidean distance	$\|\mathbf{x} - \mathbf{y}\|^2$	$\|\mathbf{x}\|^2$
Generalized I-divergence	$\sum_{i=1}^d x_i \log(\frac{x_i}{y_i}) - \sum_{i=1}^d (x_i - y_i)$	$\sum_{i=1}^d x_i \log(x_i)$
Logistic loss	$x \log(\frac{x}{y}) + (1-x) \log(\frac{1-x}{1-y})$	$x \log(x) + (1-x) \log(1-x)$
Itakura-Saito distance	$\frac{x}{y} - \log xy - 1$	$-\log x$
Hinge loss	$\max\{0, -2\mathrm{sign}(-y)x\}$	$|x|$
KL-divergence	$\sum_{i=1}^d x_i \log(\frac{x_i}{y_i})$	$\sum_{i=1}^d x_i \log(x_i)$
Mahalanobis distance	$(\mathbf{x} - \mathbf{y})^T \mathbf{A} (\mathbf{x} - \mathbf{y})$	$\mathbf{x}^T \mathbf{A} \mathbf{x}$

partitioning approaches, which may cause very expensive computation on a huge matrix.

Based on the above matrix representation, the distance between two graphs in Equation 3.1 can be formulated as the distances between a set of matrices and a set of matrix products. For example, for the two graphs shown in Figure 3.1,

$$\mathfrak{D}(G, G^s) = D(A^{(12)}, C^{(1)} B^{(12)} (C^{(2)})^T);$$

for the two graphs shown in Figure 3.2,

$$\mathfrak{D}(G, G^s) = D(A^{(12)}, C^{(1)} B^{(12)} (C^{(2)})^T) + D(A^{(13)}, C^{(1)} B^{(13)} (C^{(3)})^T).$$

Hence, finding the RSN defined in Definition 3.1 is equivalent to the following optimization problem of matrix approximation (for convenience, we assume that there exists $A^{(ij)}$ for $1 \le i < j \le m$, i.e., every pair of V_i and V_j are related to each other in G).

DEFINITION 9.1   *Given a distance function D , a set of matrices $\{A^{(ij)} \in \mathbb{R}^{n_i \times n_j}\}_{1 \le i < j \le m}$ representing a k-partite heterogeneous relation graph G, and m positive integers, $k_1, \dots, k_m$, the RSN G^s represented by $\{C^{(i)} \in \{0, 1\}^{n_i \times k_i}\}_{1 \le i \le m}$ and $\{B^{(ij)} \in \mathbb{R}^{k_i \times k_j}\}_{1 \le i < j \le m}$ is given by the minimization of*

$$L = \sum_{1 \le i < j \le m} D(A^{(ij)}, C^{(i)} B^{(ij)} (C^{(j)})^T), \qquad (9.1)$$

subject to $\sum_{k=1}^{k_i} C^{(i)}_{hk} = 1$ for $1 \le h \le n_i$.

In the above definition, the constraint on $C^{(i)}$ is to restrict $C^{(i)}$ to be an *indicator matrix*, in which each row is an indicator vector. In the definition, the distance between two matrices $D(X,Y)$ denotes the sum of the distances between each pair of elements, i.e., $D(X,Y) = \sum_{h,l} D(X_{hl}, Y_{hl})$.

For the optimization problem in Definition 3.1 or Definition 9.1, there are many choices of distance functions, which imply the different assumptions about the distribution of the weights of the edges in the given k-partite heterogeneous relation graph. For example, by using Euclidean distance function, we implicitly assume the normal distribution for the weights of the edges. Presumably for a specific distance function used in Definition 9.1, we need to derive a specific algorithm. However, a large number of useful distance functions, such as Euclidean distance, generalized I-divergence, and KL-divergence, can be generalized as the Bregman divergences [12, 110]. Based on the properties of Bregman divergences, we derive a general algorithm to minimize the objective function in Equation 9.1 under all the Bregman divergences. Table 9.1 shows a list of Bregman divergences and their corresponding Bregman convex functions. Note that Bregman divergences are nonnegative. The definition of a Bregman divergence is given as follows.

DEFINITION 9.2   *Given a strictly convex function, $\phi : S \mapsto \mathbb{R}$, defined on a convex set $S \subseteq \mathbb{R}^d$ and differentiable on the interior of S, int(S), the Bregman divergence $D_\phi : S \times \text{int}(S) \mapsto [0, \infty)$ is defined as*

$$D_\phi(x, y) = \phi(x) - \phi(y) - (x - y)^T \nabla \phi(y), \qquad (9.2)$$

where $\nabla \phi$ is the gradient of ϕ.

We prove the following theorem which is the basis of our algorithm.

THEOREM 9.1
Assume that D in Definition 9.1 is a Bregman divergence D_ϕ. If $\{C^{(i)}\}_{1 \leq i \leq m}$ and $\{B^{(ij)}\}_{1 \leq i < j \leq m}$ are an optimal solution to the minimization in Definition 9.1, then

$$(C^{(i)})^T (C^{(i)} B^{(ij)} (C^{(j)})^T - A^{(ij)}) C^{(j)} = 0 \qquad (9.3)$$

for $1 \leq i < j \leq m$.

PROOF   For convenience, we use Y to denote $C^{(i)} B^{(ij)} (C^{(j)})^T$, $\zeta(x)$ to denote $\nabla \phi(x)$, and $\xi(x)$ to denote $\nabla^2 \phi(x)$.

We compute the gradient $\nabla_{B^{(ij)}} L$, where $1 \leq i < j \leq m$ and L denotes the objective function in Equation 9.1. Using the fact that $\partial Y_{hl} / \partial B_{pq}^{(ij)} = C_{hp}^{(i)} C_{lq}^{(j)}$, we see that $\partial L / \partial B_{pq}^{(ij)}$ is given by

$$\frac{\partial}{\partial B_{pq}^{(ij)}} \{ \sum_{h,l} \phi(A_{hl}^{(ij)}) - \phi(Y_{hl}) - (A_{hl}^{(ij)} - Y_{hl}) \zeta(Y_{hl}) \}$$

$$= \sum_{h,l} -\zeta(Y_{hl}) C_{hp}^{(i)} C_{lq}^{(j)} - A_{hl}^{(ij)} \xi(Y_{hl}) C_{hp}^{(i)} C_{lq}^{(j)}$$

$$+ C_{hp}^{(i)} C_{lq}^{(j)} \zeta(Y_{hl}) + Y_{hl} \xi(Y_{hl}) C_{hp}^{(i)} C_{lq}^{(j)}$$

$$= \sum_{h,l} \xi(Y_{hl}) (Y_{hl} C_{hp}^{(i)} C_{lq}^{(j)} - A_{hl}^{(ij)} C_{hp}^{(i)} C_{lq}^{(j)}) \tag{9.4}$$

$$= [(C^{(i)})^T (\xi(Y) \odot (Y - A^{(ij)})) C^{(j)}]_{pq} \tag{9.5}$$

where $\odot$ denotes the Hadamard product or entrywise product of two matrices. By Equation (9.5), we have

$$\nabla_{B^{(ij)}} L = (C^{(i)})^T (\xi(Y) \odot (Y - A^{(ij)})) C^{(j)} \tag{9.6}$$

According to the KKT conditions, an optimal solution to Definition 9.1 satisfies $\nabla_{B^{(ij)}} L = 0$, which leads to

$$(C^{(i)})^T (\xi(Y) \odot (Y - A^{(ij)})) C^{(j)} = 0. \tag{9.7}$$

According to Definition 9.2, ϕ is strictly convex, hence, $[\xi(Y)]_{pq} > 0$ for $1 \leq p \leq k_i$ and $1 \leq q \leq k_j$. Therefore, $\xi(Y)$ can be canceled from Equation 9.7 to obtain

$$(C^{(i)})^T (Y - A^{(ij)}) C^{(j)} = 0. \tag{9.8}$$

This completes the proof of the theorem. □

The most interesting observation about Theorem 9.1 is that Equation 9.3 does not involve the distance function D_ϕ.

We propose an iterative algorithm to find a local optimal RSN represented by $\{C^{(i)}\}_{1 \leq i \leq m}$ and $\{B^{(ij)}\}_{1 \leq i < j \leq m}$ for a given k-partite heterogeneous relation graph. At each iterative step, we update one of $\{C^{(i)}\}_{1 \leq i \leq m}$ or one of $\{B^{(ij)}\}_{1 \leq i < j \leq m}$ by fixing all the others.

Since $C^{(i)}$ is an indicator matrix, we adopt the reassignment procedure such as in the k-means algorithm to update $C^{(i)}$. To determine which element of the hth row of $C^{(i)}$ is equal to 1, for $l = 1, \ldots, k_i$, we let $C_{hl}^{(i)} = 1$ and compute the objective function L in Equation 9.1 for each l, which is denoted as L_l, then

$$C_{hl^*}^{(i)} = 1 \text{ for } l^* = \arg \min_l L_l. \tag{9.9}$$

The updating rule in Equation 9.9 is equivalent to updating the edges between V_i and S_i in G^s by connecting v_{ih} to each hidden nodes in S_i to find which hidden node gives the smallest values for $D_\phi(G, G^s)$, i.e.,

$$e^s(v_{ih}, s_{il^*}) = 1 \text{ for } l^* = \arg \min_l D_\phi(G, G_l^s). \tag{9.10}$$

Algorithm 2 Relation summary network with Bregman divergences

---

Input: A k-partite heterogeneous relation graph $G = (V_1, \ldots, V_m, E)$, a Bregman divergence function D_ϕ, and m positive integers, $k_1, \ldots, k_m$.
Output: An RSN $G^s = (V_1, \ldots, V_m, S_1, \ldots, S_m, E^s)$.
Method:

  1: Initialize G^s.
  2: **repeat**
  3:   **for** $i = 1$ to m **do**
  4:     Update the edges between V_i and S_i according to Equation 9.10.
  5:   **end for**
  6:   **for** each pair of $S_i \sim S_j$ where $1 \le i < j \le m$ **do**
  7:     Update the edges between S_i and S_j according to Equation 9.12.
  8:   **end for**
  9: **until** convergence

---

where G^s_l denotes the RSN with s_{il} connecting to v_{ih}. Note that the computation for this updating involves only edges between v_{ih} and the related nodes, not all the edges.

Based on Equation 9.3 in Theorem 9.1, after a little algebraic manipulation, we have the following updating rule for each $B^{(ij)}$:

$$B^{(ij)} = ((C^{(i)})^T C^{(i)})^{-1} (C^{(i)})^T A^{(ij)} C^{(j)} ((C^{(j)})^T C^{(j)})^{-1}. \qquad (9.11)$$

This updating rule does not really involve computing inverse matrices, since $(C^{(i)})^T C^{(i)}$ is a special diagonal matrix such that $[(C^{(i)})^T C^{(i)}]_{pp} = \sum_{h=1}^{n_i} C^{(i)}_{hp}$, i.e., the number of instance nodes associated with the hidden node s_{ip}, and similarly for $(C^{(j)})^T C^{(j)}$. The updating rule in Equation 9.11 is equivalent to updating the edges between S_i and S_j in G^s by re-computing the weight of the edge between a pair of hidden nodes $s_{ip} \in S_i$ and $s_{jq} \in S_j$ as follows:

$$e^s(s_{ip}, s_{jq}) = \frac{1}{|\mathcal{U}| * |\mathcal{Z}|} \sum_{v_{ih} \in \mathcal{U}, v_{jl} \in \mathcal{Z}} e(v_{ih}, v_{jl}), \qquad (9.12)$$

where $\mathcal{U} = \{v_{ih} : e^s(v_{ih}, s_{ip}) = 1\}$, i.e., the instance nodes associated with s_{ip}; $\mathcal{Z} = \{v_{jl} : e^s(v_{jl}, s_{jq}) = 1\}$, i.e., the instance nodes associated with s_{jq}, $1 \le p \le k_i, 1 \le q \le k_j, 1 \le h \le n_i$, and $1 \le l \le n_j$. This updating rule is consistent with our intuition about the edge between two hidden nodes; i.e., it is the "summary relation" for two sets of instance nodes. It is, however, a surprising observation that the updating does not involve the distance function, i.e., this simple updating rule holds for all Bregman divergences.

The algorithm, Relation Summary Network with Bregman Divergences (RSN-BD), is summarized in Algorithm 2. RSN-BD iteratively updates the cluster structures for different types of instance nodes and summary relations among the hidden nodes. Through the hidden nodes, the cluster structures of different types of instance nodes interact with each other directly or indirectly. The interactions lead to an implicit adaptive feature reduction for each type of instance nodes which overcomes the typical high dimensionality and sparsity. RSN-BD is applicable to a wide range of problems, since it does not have restrictions on the structures of the input k-partite heterogeneous relation graph. Furthermore, the graphs from different applications may have different probabilistic distributions on their edges; it is easy for RSN-BD to adapt to this situation by simply using different Bregman divergences, since Bregman divergences correspond to a large family of exponential distributions including most common distributions, such as normal, multinomial, and Poisson distributions [33].

Note that to avoid clutter, we do not consider weighting different types of edges during the derivation. Nevertheless, it is easy to extend the proposed model and algorithm to the weighted versions.

If we assume that the number of the pairs of $V_i \sim V_j$ is $\Theta(m)$ which is typical in real applications, and let $n = \Theta(n_i)$ and $k = \Theta(k_i)$, the computational complexity of RSN-BD can be shown to be $O(tmn^2k)$ for t iterations. If we apply the k-means algorithm to each type of nodes individually by transforming the relations into features for each type of nodes, the total computational complexity is also $O(tmn^2k)$. Hence, RSN-BD is as efficient as k-means. If the edges in the graph are very sparse, the computational complexity of RSN-BD can be reduced to $O(tmrk)$ where we assume that the number of the edges between each pair of V_i and V_j is $\Theta(r)$.

Equation 9.3 in Theorem 9.1 is a necessary condition for an optimal solution, but not sufficient for the correctness of the RSN-BD algorithm. The following theorems guarantee the convergence of RSN-BD.

LEMMA 9.1

Given a Bregman divergence $D_\phi : S \times \text{int}(S) \mapsto [0, \infty)$, $A \in \mathbb{R}^{n_1 \times n_2}$ and two indicator matrices, $C^{(1)} \in \{0,1\}^{n_1 \times k_1}$ and $C^{(2)} \in \{0,1\}^{n_2 \times k_2}$, let

$$B^* = ((C^{(1)})^T C^{(1)})^{-1}(C^{(1)})^T A C^{(2)}((C^{(2)})^T C^{(2)})^{-1} \qquad (9.13)$$

then for any $B \in \mathbb{R}^{k_1 \times k_2}$,

$$D_\phi(A, C^{(1)}B(C^{(2)})^T) - D_\phi(A, C^{(1)}B^*(C^{(2)})^T) \geq 0. \qquad (9.14)$$

PROOF   For convenience we use Y to denote $C^{(1)}B(C^{(2)})^T$, Y^* to denote $C^{(1)}B^*(C^{(2)})^T$, and $\zeta(x)$ to denote $\nabla\phi(x)$. Let $\mathfrak{J}$ denote the left-hand side of

Equation 9.14.

$$\mathfrak{I} = \sum_{h,l} D_\phi(A_{hl}, Y_{hl}) - \sum_{h,l} D_\phi(A_{hl}, Y_{hl}^*)$$

$$= \sum_{h,l} \{\phi(Y_{hl}^*) - \phi(Y_{hl}) - (A_{hl} - Y_{hl})\zeta(Y_{hl})$$

$$+ (A_{hl} - Y_{hl}^*)\zeta(Y_{hl}^*)\}$$

$$= \sum_{h,l} \{\phi(Y_{hl}^*) - \phi(Y_{hl}) - [A \odot \zeta(Y)]_{hl} + [Y \odot \zeta(Y)]_{hl}$$

$$+ [A \odot \zeta(Y^*)]_{hl} - [Y^* \odot \zeta(Y^*)]_{hl}\}$$

$$= \sum_{h,l} \{\phi(Y_{hl}^*) - \phi(Y_{hl}) - [Y^* \odot \zeta(Y)]_{hl} + [Y \odot \zeta(Y)]_{hl}\}$$

$$= D_\phi(Y^*, Y)$$

$$\geq 0.$$

During the above deduction, the second and fifth equalities follow the definition of the Bregman divergences; the fourth equality follows the fact that $\sum_{h,l}[A \odot \zeta(Y^*)]_{hl} = \sum_{h,l}[Y^* \odot \zeta(Y^*)]_{hl}$ and $\sum_{h,l}[A \odot \zeta(Y)]_{hl} = \sum_{h,l}[Y^* \odot \zeta(Y)]_{hl}$ resulting from the special structure of the indicator matrix; the last inequality follows the nonnegativity of Bregman divergences. □

THEOREM 9.2
The RSN-BD algorithm (Algorithm 2) monotonically decreases the objective function in Equation 3.1.

PROOF Proving the theorem is equivalent to proving that the updating rules in Equations 9.9 and 9.11 monotonically decrease the objective function in Equation 9.1. Let $L_{(t)}$ denote the objective value after the tth iteration.

$$L_{(t)} = \sum_{1 \leq i < j \leq m} D_\phi(A^{(ij)}, C_{(t)}^{(i)} B_{(t)}^{(ij)} (C_{(t)}^{(j)})^T)$$

$$\geq \sum_{1 \leq i < j \leq m} D_\phi(A^{(ij)}, C_{(t+1)}^{(i)} B_{(t)}^{(ij)} (C_{(t+1)}^{(j)})^T)$$

$$\geq \sum_{1 \leq i < j \leq m} D_\phi(A^{(ij)}, C_{(t+1)}^{(i)} B_{(t+1)}^{(ij)} (C_{(t+1)}^{(j)})^T)$$

$$= L_{(t+1)}$$

where the first inequality follows trivially the criteria used for reassignment in Equation 9.9, and the second inequality follows Equation 9.11 and Lemma 9.1. □

Based on Theorem 9.2 and the fact that the objective function in Equation 9.1 has the lower bound 0 for a Bregman divergence, the convergence of RSN-BD is proved.

9.2　A Unified View to Clustering Approaches

In this section, we discuss the connections between the existing clustering approaches and the RSN model. By considering them as special cases or variations of the RSN model, we show that RSN provides a unified view to the existing clustering approaches.

9.2.1　Bipartite Spectral Graph Partitioning

Bipartite Spectral Graph Partitioning (BSGP) [43, 69] uses the spectral approach to partitioning a bipartite graph to find cluster structures for two types of interrelated data objects, such as words and documents. The objective function of BSGP is the normalized cut on the bipartite graph, whose affinity matrix is $\begin{bmatrix} 0 & A \\ A^T & 0 \end{bmatrix}$. After the deduction, the spectral partitioning on the bipartite graph is converted to a singular value decomposition (SVD) [43, 69].

As a graph partitioning approach, BSGP has the restriction that the clusters of different types of nodes have one-to-one associations. Under the RSN model, this restriction is equivalent to letting a hidden node connect with one and only one hidden node. Hence, the affinity matrix representing the edges between two sets of hidden nodes is restricted to a diagonal matrix. The objective function in Equation 9.1 can be formulated as

$$L = ||A^{(12)} - C^{(1)}B^{(12)}(C^{(2)})^T||^2, \tag{9.15}$$

where $|| \cdot ||$ denotes Frobenius norm, i.e., the Euclidean distance function is adopted, and A may be normalized as described in [43]. Based on this objective function, if we relax $C^{(1)}$ and $C^{(2)}$ to any othornormal matrices as in [43, 69], it immediately follows the standard result of linear algebra [58] that the minimization of L in Equation 9.15 with the diagonal constraint on B is equivalent to partial SVD. Therefore, the RSN model based on Euclidean distance function provides a simple way to understand BSGP. Comparing with BSGP, RSN-BD is more flexible to exploit the cluster structures from a bipartite graph, since it does not have one-to-one association as a constraint and is capable of adopting different distance functions.

9.2.2　Binary Data Clustering with Feature Reduction

In [85], a model is proposed to cluster binary data by clustering data points and features simultaneously, i.e., clustering with feature reduction. If we consider data points and features as two different types of nodes in a bipartite graph and the binary elements of the data matrix denote whether there exists a link between a pair of nodes, then this model is equivalent to the RSN model on a bipartite graph with unit weight edges. The objective function of this model is given in [85] as

$$O(A, X, B) = ||W - AXB^T||^2, \tag{9.16}$$

where W denotes the data matrix, A and B denote cluster memberships for data points and features, respectively, and X represents the associations between the data clusters and the feature clusters. We can see that this objective function is exactly the same as the objective function in Equation 9.1 on bipartite graph with Euclidean distance.

The immediate benefit of establishing the connection between the model proposed in [85] and the RSN model is a new solution to binary data clustering with feature reduction. In [85], the model is based on Euclidean distance. Euclidean distance function has a very wide applicability, since it implies the normal distribution and most data with a large sample size tend to have a normal distribution. However, since Bernoulli distribution is a more intuitive choice for the binary data, RSN-BD directly provides a new algorithm for clustering binary data with feature reduction by using logistic distance function (see Table 9.1), which corresponds to Bernoulli distribution.

9.2.3 Information-Theoretic Co-Clustering

[44] proposes a novel theoretic formulation to view the contingency table as an empirical joint probability distribution of two discrete random variables and develops the co-clustering algorithm, Information-Theoretic Co-clustering (ITCC), to maximize the mutual information between the clustered random variables subject to the constraints on the number of row and column clusters. Let X and Y be discrete random variables that take values in the sets $\{x_1, \ldots, x_{n_1}\}$ and $\{y_1, \ldots, y_{n_2}\}$, respectively, and $\hat{X}$ and $\hat{Y}$ be the cluster random variables that take values in the sets $\{\hat{x}_1, \ldots, \hat{x}_{k_1}\}$ and $\{\hat{y}_1, \ldots, \hat{y}_{k_2}\}$, respectively; then the objective function of ITCC is the loss in mutual information, $I(X, Y) - I(\hat{X}, \hat{Y})$.

The joint distribution of X and Y can be formulated as a bipartite graph by assigning the probability $p(x_h, y_l)$ to the weight of the edge between $v_{1h} \in V_1$ and $v_{2l} \in V_2$. If we modify Condition 1 in Definition 3.1 such that an instance node v_{ih} is connected to one and only one hidden node s_{ip} with weight $\frac{1}{\#s_{ip}}$ where $\#s_{ip}$ is the number of the instance nodes connected to s_{ip}, then in the RSN of aforementioned bipartite graph, $e^s(v_{1h}, s_{1p})$ and $e^s(v_{2l}, s_{2q})$ can be considered as $p(x_h|\hat{x}_p)$ and $p(x_l|\hat{x}_q)$, respectively; $e^s(s_{1p}, s_{2q})$ can be considered as $p(\hat{x}_p, \hat{x}_q)$. Based on this formulation, it is easy to verify that the objective function of RSN with KL-divergence is equivalent to $I(X; Y) - I(\hat{X}, \hat{Y})$. This connection between the ITCC and a variation of RSN model

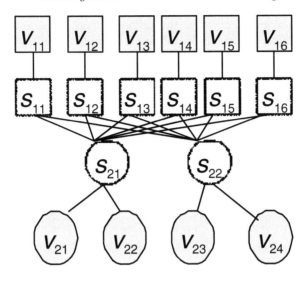

FIGURE 9.1: An RSN equivalent to k-means.

implies that the ITCC algorithm may be extended to more general cases of more than two random variables and with other loss functions.

9.2.4 K-Means Clustering

Due to its simplicity, efficiency, and broad applicability, k-means algorithm has become one of the most popular clustering algorithms. Figure 9.1 explains the relation between the RSN model and k-means. If we consider data points and features as two different types of nodes, V_2 and V_1, in a bipartite graph, and restrict feature nodes to have one-to-one associations to their hidden nodes with unit weight, then the objective function in Equation 9.1 is given as $L = ||A^{(12)} - C^{(1)}B^{(12)}(C^{(2)})^T||^2$ where $C^{(1)}$ is restricted to an identity matrix. Hence, the objective function is reduced to $L = ||A^{(12)} - B^{(12)}(C^{(2)})^T||^2$, which is exactly the matrix representation for the objective function of the k-means algorithm [138]. From Figure 9.1, we also see that since the number of feature nodes is equal to the number of their hidden nodes, k-means does not do feature reduction. Finally, we may consider RSN-BD as a generalization of k-means on k-partite heterogeneous relation graphs with various Bregman divergences and expect that it inherits the simplicity and efficiency of k-means and has a much broader applicability.

There are more clustering approaches in the literature that may be considered as the special cases or variations of the RSN model. For example, the subspace clustering [3], which clusters the data points in a high-dimensional space around a different subset of the dimensions, can be considered as an extension of Figure 9.1 such that s_{21} or s_{22} only connects to a subset of S_1. Spectral relational clustering [87] can be considered as using the spectral approach to solve the RSN model under Euclidean distance.

By examining the connections between existing clustering approaches and the RSN model, we conclude that the RSN model provides a unified view to the existing clustering approaches. Moreover, the idea of RSN is more general than the proposed model based on Definition 3.1 and Equation 3.1. For example, if we change the definition of distance between graphs in Equation 3.1, we may find totally different ways to mine hidden structures from a k-partite heterogeneous relation graph, and as a result, we may obtain new variations for existing clustering approaches.

Chapter 10

Homogeneous Relational Data Clustering

In Chapter 4, we proposed a general model based on the graph approximation to learn relation-pattern-based cluster structures from a graph. The model generalizes the traditional graph partitioning approaches and is applicable to learning the various cluster structures. In this chapter, under the model we derive a family of algorithms which are flexible to learn various cluster structures and easy to incorporate the prior knowledge of the cluster structures. Specifically, we derive algorithms for the basic CLGA model in Definition 4.1 and its extensions, soft CLGA model and balanced CLGA model.

10.1   Hard CLGA Algorithm

The CLGA model in Definition 4.1 seeks hard cluster membership for each node and the problem can be shown to be NP-hard. The proof is easy since based on Theorem 11.4 it can be reduced to the graph partitioning problem, which is NP-hard. We derive an alternative optimization algorithm for the hard CLGA model.

We prove the following theorem which is the basis of our algorithm.

THEOREM 10.1
If $C \in \{0,1\}^{n \times k}$ and $B \in \mathbb{R}_+^{k \times k}$ are an optimal solution to the minimization in Definition 4.1, then

$$B = (C^T C)^{-1} C^T A C (C^T C)^{-1}. \qquad (10.1)$$

PROOF   The objective function in Definition 4.1 can be expanded as follows:

$$\begin{aligned}
L &= ||A - CBC^T||^2 \\
&= \mathrm{tr}((A - CBC^T)^T (A - CBC^T)) \\
&= \mathrm{tr}(A^T A) - 2\mathrm{tr}(CBC^T A) - \mathrm{tr}(CBC^T CBC^T)
\end{aligned}$$

Algorithm 3 Hard CLGA algorithm

Input: A graph affinity matrix A and a positive integer k.
Output: A cluster membership matrix C and a cluster structure matrix B.
Method:

1: Initialize B.
2: **repeat**
3:    **for** $h = 1$ to n **do**
4:        $C_{hp^*} = 1$ for $p^* = \arg\min_p L_p$ where L_p denotes $||A - CBC^T||^2$ for $C_{hp} = 1$.
5:    **end for**
6:    $B = (C^T C)^{-1} C^T A C (C^T C)^{-1}$.
7: **until** convergence

Take the derivative with respect to B, we obtain

$$\frac{\partial L}{\partial B} = -2C^T AC + 2C^T CBC^T C. \tag{10.2}$$

Solve $\frac{\partial L}{\partial B} = 0$ to obtain

$$B = (C^T C)^{-1} C^T A C (C^T C)^{-1}. \tag{10.3}$$

Note that $C^T C$ is a special diagonal matrix such that $[C^T C]_{pp} = |\pi_p|$, the size of the pth cluster, and since A is a nonnegative symmetric matrix, and so is B. This completes the proof of the theorem. $\qquad\square$

Based on Theorem 10.1, we propose an iterative algorithm which alternatively updates B and C and converges to a local optimum. First, we fix C and update B. Eq 10.1 in Theorem 10.1 provides an updating rule for B. This updating rule can be implemented more efficiently than it looks like. First, it does not really involve computing inverse matrices, since $C^T C$ is a special diagonal matrix with the sizes of the clusters on its diagonal; second, the product of $C^T AC$ can be calculated without normal matrix multiplication, since C is an indicator matrix.

Second, we fix B and update C. Since each row of C is an indicator vector with only one element equal to 1, we adopt the reassignment procedure to update C row by row. To determine which element of the hth row of C is equal to 1, for $p = 1, \ldots, k$, each time we let $C_{hp} = 1$ and compute the objective function $L = ||A - CBC^T||^2$, which is denoted as L_p, then

$$C_{hp^*} = 1 \text{ for } p^* = \arg\min_p L_p. \tag{10.4}$$

Note that when we update the hth row of C, the necessary computation involves only the hth row or column of A and CBC^T.

The algorithm, hard CLGA, is summarized in Algorithm 3. The hard CLGA algorithm learns the general cluster structures from a graph, since it does not put any constraint on B. However, it is trivial to modify it to solve the variations of the general CLGA model, such as IGP, GGP, IBCL, and GBCL in Table 4.1. For example, the hard IGP algorithm works as follows: fix B as the identity matrix and simply update C by updating rule Equation 10.4 until convergence; the hard GGP algorithm works as follows: fix the off-diagonal elements of B as zero; update the diagonal elements of B by updating rule Equation 10.1 and update C by updating rule Equation 10.4 until convergence. Besides the variations in Table 4.1, it is easy for Algorithm 3 to incorporate other prior knowledge through B.

The complexity of hard CLGA can be shown to be $O(tn^2k)$ where t is the number of iterations. It can be further reduced for sparse graphs. When applied to graph partitioning task, the hard CLGA algorithm is computationally more efficient than the popular spectral approaches which involve expensive eigenvector computation and extra postprocessing on eigenvectors to obtain the partitioning. Compared with the multilevel approaches such as METIS [75], CLGA does not restrict clusters to have an equal size.

The proof of the convergence of Algorithm 3 is easy due to the following facts. First, based on Theorem 10.1, the objective function is nonincreasing under updating rule Equation 10.1; second, by the criteria for reassignment in updating rule Equation 10.4, it is trivial to show that the objective function is nonincreasing under updating rule Equation 10.4.

10.2   Soft CLGA Algorithm

In the hard CLGA model of Definition 4.1, each node belongs to only one cluster. It is natural to extend it to a soft version in which each node could belong to more than one cluster with certain degrees. Formally, we define the soft CLGA model as follows.

DEFINITION 10.1   *Given an undirected graph $G = (\mathcal{V}, \mathcal{E}, A)$ where $A \in \mathbb{R}_+^{n \times n}$ is the affinity matrix, and a positive integer k, the optimized clusters are given by the minimization,*

$$\min_{\substack{C \in \mathbb{R}_+^{n \times k}, B \in \mathbb{R}_+^{k \times k} \\ C1=1}} ||A - CBC^T||^2. \tag{10.5}$$

In the soft CLGA model, C is a soft membership matrix such that C_{ij} denotes the degree that the ith node is associated with the jth cluster and the sum of all the degrees for each node equals to 1. Since the soft CLGA model defines a constrained nonconvex optimization, it is not realistic to expect an algorithm to find a global optimum. We propose an alternative optimization algorithm which converges to a local optimum.

Deriving the updating rules for C and B in the soft CLGA model is more difficult than that in the hard CLGA model. First, it is difficult to deal with the constraint $\sum_j C_{ij} = 1$ efficiently. Hence, we transform it to a "soft" constraint i.e., we implicitly enforce the constraint by adding a penalty term, $\alpha ||C\mathbf{1} - \mathbf{1}||^2$ where $\mathbf{1}$ is a k-dimension vector consisting of 1s, and α is a positive constant. Therefore, we obtain the following optimization:

$$\min_{C \in \mathbb{R}_+^{n \times k}, B \in \mathbb{R}_+^{k \times k}} ||A - CBC^T||^2 + \alpha ||C\mathbf{1} - \mathbf{1}||^2. \tag{10.6}$$

Fixing B, the objective function in Equation 10.6 is quartic with respect to C. We derive a simple and efficient updating rule for C based on the bound optimization procedure [83, 107]. The basic idea is to construct an auxiliary function which is a convex upper bound for the original objective function based on the solution obtained from the previous iteration. Then, a new solution for the current iteration is obtained by minimizing this upper bound.

DEFINITION 10.2 $G(S, S^t)$ *is an auxiliary function for $F(S)$ if $G(S, S^t) \geq F(S)$ and $G(S, S) = F(S)$.*

The auxiliary function is useful due to the following lemma.

LEMMA 10.1
If G is an auxiliary function, then F is nonincreasing under the updating rule $S^{t+1} = \arg\min_S G(S, S^t)$.

PROOF $F(S^{t+1}) \leq G(S^{t+1}, S^t) \leq G(S^t, S^t) \leq F(S^t)$.   ⬚

We propose an auxiliary function for C in the following lemma.

LEMMA 10.2

$$G(C, \tilde{C}) = \sum_{ij}(A_{ij} + \frac{\alpha}{n} - 2\sum_{gh}(A_{ij}\tilde{C}_{ig}B_{gh}\tilde{C}_{jh}(1 + 2\log C_{jh}$$

$$-2\log \tilde{C}_{jh}) + \frac{\alpha}{nk}\tilde{C}_{jh}(1 + \log C_{jh} - \log \tilde{C}_{jh})) +$$

$$\sum_{gh}([\tilde{C}B\tilde{C}^T]_{ij}\tilde{C}_{ig}B_{gh}\tilde{C}_{jh}\frac{C_{jh}^4}{\tilde{C}_{jh}^4} +$$

$$\frac{\alpha}{2nk}[\tilde{C}\mathbf{1}]_j\tilde{C}_{jh}(\frac{C_{jh}^4}{\tilde{C}_{jh}^4}+1)))$$

is an auxiliary function for

$$F(C) = ||A - CBC^T||^2 + \alpha||C\mathbf{1} - \mathbf{1}||^2. \qquad (10.7)$$

PROOF For convenience, we let $\beta = \frac{\alpha}{nk}$.

$$F(C) = \sum_{ij}((A_{ij} - \sum_{gh}C_{ig}B_{gh}C_{jh})^2 + \beta\sum_{gh}(C_{jh}-1)^2)$$

$$\leq \sum_{ij}(\sum_{gh}\frac{\tilde{C}_{ig}B_{gh}\tilde{C}_{jh}}{[\tilde{C}B\tilde{C}^T]_{ij}}(A_{ij} - \frac{[\tilde{C}B\tilde{C}^T]_{ij}}{\tilde{C}_{ig}B_{gh}\tilde{C}_{jh}}C_{ig}B_{gh}C_{jh})^2$$

$$+\beta\sum_{gh}\frac{\tilde{C}_{jh}}{[\tilde{C}\mathbf{1}]_j}(\frac{[\tilde{C}\mathbf{1}]_j}{\tilde{C}_{jh}}C_{jh} - 1)^2)$$

$$= \sum_{ij}(A_{ij} - 2\sum_{gh}A_{ij}C_{ig}B_{gh}C_{jh} +$$

$$\sum_{gh}\frac{[\tilde{C}B\tilde{C}^T]_{ij}}{\tilde{C}_{ig}B_{gh}\tilde{C}_{jh}}C_{ig}^2B_{gh}^2C_{jh}^2 + \beta\sum_{gh}\frac{[\tilde{C}\mathbf{1}]_j}{\tilde{C}_{jh}}C_{jh}^2$$

$$-2\beta\sum_{gh}C_{jh} + k\beta)$$

$$= \sum_{ij}(A_{ij} + k\beta - 2\sum_{gh}(A_{ij}\tilde{C}_{ig}B_{gh}\tilde{C}_{jh}\frac{C_{ig}C_{jh}}{\tilde{C}_{ig}\tilde{C}_{jh}} +$$

$$\beta\tilde{C}_{jh}\frac{C_{jh}}{\tilde{C}_{jh}}) + \sum_{gh}([\tilde{C}B\tilde{C}^T]_{ij}\tilde{C}_{ig}B_{gh}\tilde{C}_{jh}\frac{C_{ig}^2C_{jh}^2}{\tilde{C}_{ig}^2\tilde{C}_{jh}^2}$$

$$+\beta[\tilde{C}\mathbf{1}]_j\tilde{C}_{jh}\frac{C_{jh}^2}{\tilde{C}_{jh}^2}))$$

$$\leq \sum_{ij}(A_{ij} + k\beta - 2\sum_{gh}(A_{ij}\tilde{C}_{ig}B_{gh}\tilde{C}_{jh}(1 + \log C_{ig}$$

$$+ \log C_{jh} - \log\tilde{C}_{ig} - \log\tilde{C}_{jh}) + \beta\tilde{C}_{jh}(1 + \log C_{jh} -$$

$$\log\tilde{C}_{jh})) + \sum_{gh}(\frac{1}{2}[\tilde{C}B\tilde{C}^T]_{ij}\tilde{C}_{ig}B_{gh}\tilde{C}_{jh}(\frac{C_{ig}^4}{\tilde{C}_{ig}^4} + \frac{C_{jh}^4}{\tilde{C}_{jh}^4})$$

$$+\frac{1}{2}\beta[\tilde{C}\mathbf{1}]_j\tilde{C}_{jh}(\frac{C_{jh}^4}{\tilde{C}_{jh}^4} + 1)))$$

$$= \sum_{ij}(A_{ij} + k\beta - 2\sum_{gh}(A_{ij}\tilde{C}_{ig}B_{gh}\tilde{C}_{jh}(1 + 2\log C_{jh}$$

$$-2 \log \tilde{C}_{jh}) + \beta \tilde{C}_{jh}(1 + \log C_{jh} - \log \tilde{C}_{jh})) +$$

$$\sum_{gh}([\tilde{C}B\tilde{C}^T]_{ij}\tilde{C}_{ig}B_{gh}\tilde{C}_{jh}\frac{C_{jh}^4}{\tilde{C}_{jh}^4} +$$

$$\frac{1}{2}\beta[\tilde{C}\mathbf{1}]_j\tilde{C}_{jh}(\frac{C_{jh}^4}{\tilde{C}_{jh}^4} + 1))).$$

During the above deduction, the second step uses Jensen's inequality and the fifth step uses the inequalities $x \geq 1 + \log x$ and $x^2 + y^2 \geq 2xy$. □

The following theorem provides the updating rule for C.

THEOREM 10.2
The objective function $F(C)$ in Equation 10.7 is nonincreasing under the updating rule,

$$C = \tilde{C} \odot (\frac{A\tilde{C}B + \frac{\alpha}{2}}{\tilde{C}B\tilde{C}^T\tilde{C}B + \frac{\alpha}{2}\tilde{C}E})^{\frac{1}{4}} \tag{10.8}$$

where $\tilde{C}$ denotes the solution from the previous iteration, E denotes a $k \times k$ matrix of 1s, $\odot$ denotes entrywise product, and the division between two matrix is entrywise division.

The theorem can be proven by solving $\frac{\partial G(C,\tilde{C})}{\partial C_{jh}} = 0$ and using Lemma 10.1.

Similarly, we present the following theorems to derive updating rules for B. Note that Theorem 10.1 cannot be used to update B, since C does not have the special structure of the indicator matrix in this case; updating rule Equation 10.1 cannot guarantee that B is nonnegative. To guarantee that B can be updated appropriately, we have the following theorems.

LEMMA 10.3

$$G(B, \tilde{B}) = \sum_{ij}(A_{ij} - 2\sum_{gh}A_{ij}C_{ig}B_{gh}C_{jh} +$$

$$\sum_{gh}[C\tilde{B}C]_{ij}C_{ig}C_{jh}\frac{B_{gh}^2}{\tilde{B}_{gh}})$$

is an auxiliary function for

$$F(B) = ||A - CBC^T||^2. \tag{10.9}$$

Algorithm 4 Soft CLGA algorithm

Input: A graph affinity matrix A and a positive integer k.
Output: A cluster membership matrix C and a cluster structure matrix B.
Method:

1: Initialize B and C.
2: **repeat**
3:
$$B = B \odot \frac{C^T A C}{C^T C B C^T C}.$$

4:
$$C = C \odot \left(\frac{ACB + \frac{\alpha}{2}}{CBC^T CB + \frac{\alpha}{2}CE}\right)^{\frac{1}{4}}$$

5: **until** convergence

THEOREM 10.3

The objective function $F(B)$ in Equation 10.9 is nonincreasing under the updating rule

$$B = \tilde{B} \odot \frac{C^T A C}{C^T C \tilde{B} C^T C}. \tag{10.10}$$

Following the way to prove Lemma 10.2 and Theorem 10.2, it is easy to prove the above theorems. We omit details here.

The soft CLGA algorithm is summarized in Algorithm 4. The implementation of the algorithm is simple and it is easy to take the advantage of the distributed computation for very large data. The complexity of the algorithm is still $O(tn^2k)$ for t iterations and it can be further reduced for sparse graphs. The convergence of the soft CLGA algorithm is guaranteed by Theorems 10.2 and 10.3.

Like the hard CLGA algorithm, the soft CLGA can be applied to learning the specific types of cluster structures by enforcing the corresponding constraint on B. For example, soft IGP and soft GGP provide another two new graph partitioning algorithms and they deal with the soft graph partitioning problem which has not been addressed extensively in the literature of graph partitioning.

10.3 Balanced CLGA Algorithm

In some applications, users may be interested in clusters of a balanced size. We propose the balanced CLGA model as follows.

DEFINITION 10.3   *Given an undirected graph $G = (\mathcal{V}, \mathcal{E}, A)$ where $A \in \mathbb{R}_+^{n \times n}$ is the affinity matrix, and a positive integer k, the optimized clusters are given by the minimization,*

$$\min_{\substack{C \in \mathbb{R}_+^{n \times k}, B \in \mathbb{R}_+^{k \times k} \\ C^T \mathbf{1} = \mathbf{1}}} ||A - CBC^T||^2. \tag{10.11}$$

Unlike $C\mathbf{1} = \mathbf{1}$ in the soft CLGA model, we have $C^T\mathbf{1} = \mathbf{1}$ in the balanced CLGA model, i.e., the sum of the elements in each column of the cluster membership matrix equals to 1. This constraint enforces that for each cluster, the sum of degrees that each node is associated with this cluster equals to 1. As a result, the more the nodes in a certain cluster, the smaller the average degree that each node is associated with this cluster, the more a node tends to belong to other clusters with a relative larger degree. Therefore, compared with the other two CLGA models, the balanced CLGA model tends to provide more balanced clusters. Note that this constraint does not enforce strictly balanced clusters of an equal size and it just pushes the model to provide clusters as balanced as possible.

Due to the following lemma, the balanced CLGA model can be simplified by dropping the constraint $C^T\mathbf{1} = \mathbf{1}$.

LEMMA 10.4
If $C \in \mathbb{R}_+^{n \times k}$, $B \in \mathbb{R}_+^{k \times k}$, and $D \in \mathbb{R}_+^{k \times k}$ is a diagonal matrix s.t. $D_{jj} = \frac{1}{\sum_i C_{ij}}$, then $CBC^T = CDD^{-1}BD^{-1}(CD)^T$ and $(CD)^T\mathbf{1} = \mathbf{1}$.

PROOF   omitted.   ⬚

Lemma 10.4 implies that we can always normalize C to satisfy the constraint without changing the value of the objective function. Hence, the balanced CLGA can be reduced to the following optimization:

$$\min_{C \in \mathbb{R}_+^{n \times k}, B \in \mathbb{R}_+^{k \times k}} ||A - CBC^T||^2. \tag{10.12}$$

Following the way to derive the soft CLGA algorithm, we derive the balanced CLGA algorithm. Since Lemma 10.3 and Theorem 10.2 still hold true for the optimization in Equation 10.12, the balanced CLGA has the same updating rule for B as the soft CLGA algorithm. By dropping off the term $\alpha||C\mathbf{1} - \mathbf{1}||^2$ in Equation 10.6, the theorems similar to Lemmas 10.2 and 10.2 can be obtained to derive a simpler updating rule for C. The balanced CLGA algorithm is summarized in Algorithm 5. Similarly, by enforcing constraints on B, it is easy to obtain other versions of the balanced CLGA algorithm, such as balanced IGP and balanced GGP.

Algorithm 5 Balanced CLGA algorithm

Input: A graph affinity matrix A and a positive integer k.
Output: A cluster membership matrix C and a cluster structure matrix B.
Method:

1: Initialize B and C.
2: **repeat**
3:
$$B = B \odot \frac{C^T AC}{C^T CBC^T C}.$$

4:
$$C = C \odot \left(\frac{ACB}{CBC^T CB}\right)^{\frac{1}{4}} \qquad (10.13)$$

5: **until** convergence
6: Let D be a diagonal matrix s.t. $D_{jj} = \frac{1}{\sum_i C_{ij}}$.
7: $C = CD$
8: $B = D^{-1}BD^{-1}$

Chapter 11

General Relational Data Clustering

In this Chapter 5 we proposed a probabilistic model for relational clustering, which also provides a principal framework to unify various important clustering tasks, including traditional attributes-based clustering, semi-supervised clustering, co-clustering, and graph clustering. The proposed model seeks to identify cluster structures for each type of data objects and interaction patterns between different types of objects. In this chapter, under this model, we propose parametric hard and soft relational clustering algorithms under a large number of exponential family distributions. The algorithms are applicable to relational data of various structures and at the same time unifies a number of stat-of-the-art clustering algorithms: co-clustering algorithms, the k-partite graph clustering, and semi-supervised clustering based on hidden Markov random fields.

11.1 Mixed Membership Relational Clustering Algorithm

In this section, based on the Mixed Membership Relational Clustering Algorithm (MMRC) model we derive parametric soft and hard relational clustering algorithms under a large number of exponential family distributions.

11.1.1 MMRC with Exponential Families

To avoid clutter, instead of general relational data, we use relational data similar to the one in Figure 5.1b, which is a representative relational data set containing all three types of information for relational data: attributes, homogeneous relations, and heterogeneous relations. However, the derivation and algorithms are applicable to general relational data.

For the relational data set in Figure 5.1b, we have three types of objects: one attribute matrix F, one homogeneous relation matrix S, and one heterogeneous relation matrix R. Based on Equation 5.1, we have the following likelihood function,

$$\mathcal{L}(\Omega|\Psi) = Pr(C^{(1)}|\Lambda^{(1)})Pr(C^{(2)}|\Lambda^{(2)})Pr(F|\Theta C^{(1)})Pr(S|(C^{(1)})^T\Gamma C^{(1)})$$
$$Pr(R|(C^{(1)})^T\Upsilon C^{(2)}).$$
$$(11.1)$$

Our goal is to maximize the likelihood function in Equation 11.1 to estimate unknown parameters.

For the likelihood function in Equation 11.1, the specific forms of condition distributions for attributes and relations depend on specific applications. Presumably, for a specific likelihood function, we need to derive a specific algorithm. However, a large number of useful distributions, such as normal distribution, Poisson distribution, and Bernoulli distributions, belong to exponential families and the distribution functions of exponential families can be formulated as a general form. This nice property facilitates us to derive a general EM algorithm for the MMRC model.

It is shown in the literature [13,33] that there exists bijection between exponential families and Bregman divergences [110]. For example, the normal distribution, Bernoulli distribution, multinomial distribution, and exponential distribution correspond to Euclidean distance, logistic loss, KL-divergence, and Itakura-Satio distance, respectively. Based on the bijection, an exponential family density $Pr(x)$ can always be formulated as the following expression with a Bregman divergence D_ϕ:

$$Pr(x) = \exp(-D_\phi(x,\mu))f_\phi(x), \qquad (11.2)$$

where $f_\phi(x)$ is a uniquely determined function for each exponential probability density, and μ is the expectation parameter. Therefore, for the MMRC model under exponential family distributions, we have the following:

$$Pr(F|\Theta C^{(1)}) = \exp(-D_{\phi_1}(F,\Theta C^{(1)}))f_{\phi_1}(F) \qquad (11.3)$$

$$Pr(S|(C^{(1)})^T\Gamma C^{(1)}) = \exp(-D_{\phi_2}(S,(C^{(1)})^T\Gamma C^{(1)}))f_{\phi_2}(S) \qquad (11.4)$$

$$Pr(R|(C^{(1)})^T\Upsilon C^{(2)}) = \exp(-D_{\phi_3}(R,(C^{(1)})^T\Upsilon C^{(2)}))f_{\phi_3}(R) \qquad (11.5)$$

In the above equations, a Bregman divergence of two matrices is defined as the sum of the Bregman divergence of each pair of elements from the two matrices. Another advantage of the above formulation is that under this formulation, the parameters, Θ, Λ, and Υ, are expectations of intuitive interpretations. Θ consists of center vectors of attributes; Γ provides an intuitive summary of the cluster structures within the same type of objects, since $\Gamma_{gh}^{(1)}$ implies expectation relations between the gth cluster and the hth cluster of type 1 objects; similarly, Υ provides an intuitive summary for the cluster structures between the objects of different types. In the above formulation, we use different Bregman divergences, D_{ϕ_1}, D_{ϕ_2}, and D_{ϕ_3}, for the attributes, homogeneous relations, and heterogeneous relations, since they could have different distributions in real applications. For example, suppose that we have $\Theta^{(1)} = \begin{bmatrix} 1.1 & 2.3 \\ 1.5 & 2.5 \end{bmatrix}$ for normal distribution, $\Gamma^{(1)} = \begin{bmatrix} 0.9 & 0.1 \\ 0.1 & 0.7 \end{bmatrix}$ for Bernoulli distribution , and $\Upsilon^{(12)} = \begin{bmatrix} 1 & 3 \\ 3 & 1 \end{bmatrix}$ for Poisson distribution, then the cluster structures of the data are very intuitive. First, the center attribute vectors for the two clusters of type 1 are $\begin{bmatrix} 1.1 \\ 1.5 \end{bmatrix}$ and $\begin{bmatrix} 2.3 \\ 2.5 \end{bmatrix}$; second, by $\Gamma^{(1)}$ we know that type 1 nodes from different clusters are barely related and cluster 1 is denser than cluster 2; third, by $\Upsilon^{(12)}$ we know that cluster 1 of type 1 nodes are related to cluster 2 of type 2 nodes more strongly than to cluster 1 of type 2, and so on so forth.

Since the distributions of $C^{(1)}$ and $C^{(2)}$ are modeled as multinomial distributions, we have the following:

$$Pr(C^{(1)}|\Lambda^{(1)}) = \prod_{p=1}^{n_1} \prod_{g=1}^{k_1} (\Lambda_{gp}^{(1)})^{C_{gp}^{(1)}} \tag{11.6}$$

$$Pr(C^{(2)}|\Lambda^{(2)}) = \prod_{q=1}^{n_2} \prod_{h=1}^{k_2} (\Lambda_{hq}^{(2)})^{C_{hq}^{(2)}}. \tag{11.7}$$

Substituting Equations 11.3 through 11.7 into Equation 11.1 and taking some algebraic manipulations, we obtain the following log-likelihood function for MMRC under exponential families,

$$
\begin{aligned}
\log(\mathcal{L}(\Omega|\Psi)) = \sum_{p=1}^{n_1} \sum_{g=1}^{k_1} C_{gp}^{(1)} \log \Lambda_{gp}^{(1)} + \sum_{q=1}^{n_2} \sum_{h=1}^{k_2} C_{hq}^{(2)} \log \Lambda_{hq}^{(2)} \\
- D_{\phi_1}(F, \Theta C^{(1)}) - D_{\phi_2}(S, (C^{(1)})^T \Gamma C^{(1)}) \\
- D_{\phi_3}(R, (C^{(1)})^T \Upsilon C^{(2)}) + \tau,
\end{aligned}
\tag{11.8}
$$

where $\tau = \log f_{\phi_1}(F) + \log f_{\phi_2}(S) + \log f_{\phi_3}(R)$, which is a constant in the log-likelihood function.

Expectation Maximization (EM) is a general approach to find the maximum-likelihood estimate of the parameters when the model has latent variables. EM does the maximum likelihood estimation by iteratively maximizing the expectation of the complete (log-)likelihood, which is the following under the MMRC model:

$$Q(\Omega, \tilde{\Omega}) = E[\log(\mathcal{L}(\Omega|\Psi))|C^{(1)}, C^{(2)}, \tilde{\Omega}], \tag{11.9}$$

where $\tilde{\Omega}$ denotes the current estimation of the parameters and Ω is the new parameter set that we optimize to increase Q. Two steps, E-step and M-step, are alternatively performed to maximize the objective function in Equation 11.9.

11.1.2   Monte Carlo E-Step

In the E-step, based on Bayes' rule, the posterior probability of the latent variables,

$$Pr(C^{(1)}, C^{(2)}|F, S, R, \tilde{\Omega}) = \frac{Pr(C^{(1)}, C^{(2)}, F, S, R|\tilde{\Omega})}{\sum_{C^{(1)}, C^{(2)}} Pr(C^{(1)}, C^{(2)}, F, S, R|\tilde{\Omega})}, \quad (11.10)$$

is updated using the current estimation of the parameters. However, conditioning on observations, the latent variables are not independent, i.e., there exist dependencies between the posterior probabilities of $C^{(1)}$ and $C^{(2)}$, and between those of $C^{(1)}_{\cdot p}$ and $C^{(1)}_{\cdot q}$. Hence, directly computing the posterior based on Equation 11.10 is prohibitively expensive.

There exist several techniques for computing intractable posterior, such as Monte Carlo approaches, belief propagation, and variational methods. We follow a Monte Carlo approach, Gibbs sampler, which is a method of constructing a Markov chain whose stationary distribution is the distribution to be estimated.

It is easy to compute the posterior of a latent indicator vector while fixing all other latent indicator vectors, i.e.,

$$Pr(C^{(1)}_{\cdot p}|C^{(1)}_{\cdot -p}, C^{(2)}, F, S, R, \tilde{\Omega}) = \frac{Pr(C^{(1)}, C^{(2)}, F, S, R|\tilde{\Omega})}{\sum_{C^{(1)}_{\cdot p}} Pr(C^{(1)}, C^{(2)}, F, S, R|\tilde{\Omega})}, \quad (11.11)$$

where $C^{(1)}_{\cdot -p}$ denotes all the latent indicator vectors except for $C^{(1)}_{\cdot p}$. Therefore, we present the following Markov chain to estimate the posterior in Equation 11.10.

- Sample $C^{(1)}_{\cdot 1}$
  from distribution $Pr(C^{(1)}_{\cdot 1}|C^{(1)}_{\cdot -1}, C^{(2)}, F, S, R, \tilde{\Omega})$;

- $\cdots\cdots$

- Sample $C^{(1)}_{\cdot n_1}$
  from distribution $Pr(C^{(1)}_{\cdot n_1}|C^{(1)}_{\cdot -n_1}, C^{(2)}, F, S, R, \tilde{\Omega})$;

- Sample $C^{(2)}_{\cdot 1}$
  from distribution $Pr(C^{(2)}_{\cdot 1}|C^{(2)}_{\cdot -1}, C^{(1)}, F, S, R, \tilde{\Omega})$;

- $\cdots\cdots$

- Sample $C^{(2)}_{\cdot n_2}$
  from distribution $Pr(C^{(2)}_{\cdot n_2}|C^{(2)}_{\cdot -n_2}, C^{(1)}, F, S, R, \tilde{\Omega})$

Note that at each sampling step in the above procedure, we use the latent indicator variables sampled from previous steps. The above procedure iterates until the stop criterion is satisfied. It can be shown that the above procedure is a Markov chain converging to $Pr(C^{(1)}, C^{(2)}|F, S, R, \tilde{\Omega})$. Assume that we keep l samples for estimation; then the posterior can be obtained simply by the empirical joint distribution of $C^{(1)}$ and $C^{(2)}$ in the l samples.

11.1.3   M-Step

After the E-step, we have the posterior probability of latent variables to evaluate the expectation of the complete log-likelihood,

$$Q(\Omega, \tilde{\Omega}) = \sum_{C^{(1)}, C^{(2)}} \log(\mathcal{L}(\Omega|\Psi)) Pr(C^{(1)}, C^{(2)}|F, S, R, \tilde{\Omega}). \tag{11.12}$$

In the M-step, we optimize the unknown parameters by

$$\Omega^* = \arg\max_{\Omega} Q(\Omega, \tilde{\Omega}). \tag{11.13}$$

First, we derive the update rules for membership parameters $\Lambda^{(1)}$ and $\Lambda^{(2)}$. To derive the expression for each $\Lambda_{hp}^{(1)}$, we introduce the Lagrange multiplier α with the constraint $\sum_{g=1}^{k_1} \Lambda_{gp}^{(1)} = 1$, and solve the following equation:

$$\frac{\partial}{\partial \Lambda_{hp}^{(1)}} \{Q(\Omega, \tilde{\Omega}) + \alpha(\sum_{g=1}^{k_1} \Lambda_{gp}^{(1)} - 1)\} = 0. \tag{11.14}$$

Substituting Equations 11.8 and 11.12 into Equation 11.14, after some algebraic manipulations, we have

$$Pr(C_{hp}^{(1)} = 1|F, S, R, \tilde{\Omega}) - \alpha\Lambda_{hp}^{(1)} = 0. \tag{11.15}$$

Summing both sides over h, we obtain $\alpha = 1$ resulting in the following update rule:

$$\Lambda_{hp}^{(1)} = Pr(C_{hp}^{(1)} = 1|F, S, R, \tilde{\Omega}), \tag{11.16}$$

i.e., $\Lambda_{hp}^{(1)}$ is updated as the posterior probability that the pth object is associated with the hth cluster. Similarly, we have the following update rule for $\Lambda_{hp}^{(2)}$:

$$\Lambda_{hp}^{(2)} = Pr(C_{hp}^{(2)} = 1|F, S, R, \tilde{\Omega}). \tag{11.17}$$

Second, we derive the update rule for Θ. Based on Equations 11.8 and 11.12, optimizing Θ is equivalent to the following optimization:

$$\arg\min_{\Theta} \sum_{C^{(1)}, C^{(2)}} D_{\phi_1}(F, \Theta C^{(1)}) Pr(C^{(1)}, C^{(2)}|F, S, R, \tilde{\Omega}). \tag{11.18}$$

We reformulate the above expression,

$$\arg\min_{\Theta} \sum_{C^{(1)}} \sum_{g=1}^{k_1} \sum_{p:C_{gp}^{(1)}=1} D_{\phi_1}(F_{\cdot p}, \Theta_{\cdot g}) Pr(C_{gp}^{(1)} = 1|F, S, R, \tilde{\Omega}). \tag{11.19}$$

To solve the above optimization, we make use of an important property of Bregman divergence presented in the following theorem.

THEOREM 11.1
Let X be a random variable taking values in $\mathcal{X} = \{x_i\}_{i=1}^n \subset S \subseteq \mathbb{R}^d$ following v. Given a Bregman divergence $D_\phi : S \times \text{int}(S) \mapsto [0, \infty)$, the problem

$$\min_{s \in S} E_v[D_\phi(X, s)] \tag{11.20}$$

has a unique minimizer given by $s^* = E_v[X]$.

The proof of Theorem 11.1 is omitted (please refer [13, 110]). Theorem 11.1 states that the Bregman representative of a random variable is always the expectation of the variable. Based on Theorem 11.1 and the objective function in Equation 11.19, we update $\Theta_{\cdot g}$ as follows:

$$\Theta_{\cdot g} = \frac{\sum_{p=1}^{n_1} F_{\cdot p} Pr(C_{gp}^{(1)} = 1 | F, S, R, \tilde{\Omega})}{\sum_{p=1}^{n_1} Pr(C_{gp}^{(1)} = 1 | F, S, R, \tilde{\Omega})}. \tag{11.21}$$

Third, we derive the update rule for Γ. Based on Equations 11.8 and 11.12, we formulate optimizing Γ as the following optimization,

$$\arg\min_{\Gamma} \sum_{C^{(1)}} \sum_{g=1}^{k_1} \sum_{h=1}^{k_1} \sum_{\substack{p:C_{gp}^{(1)}=1, \\ q:C_{hq}^{(1)}=1}} D_{\phi_2}(S_{pq}, \Gamma_{gh}) \tilde{p}, \tag{11.22}$$

where $\tilde{p}$ denotes $Pr(C_{gp}^{(1)} = 1, C_{hq}^{(1)} = 1 | F, S, R, \tilde{\Omega})$, and $1 \leq p, q \leq n_1$. Based on Theorem 11.1, we update each Γ_{gh} as follows:

$$\Gamma_{gh} = \frac{\sum_{p,q=1}^{n_1} S_{pq} Pr(C_{gp}^{(1)} = 1, C_{hq}^{(1)} = 1 | F, S, R, \tilde{\Omega})}{\sum_{p,q=1}^{n_1} Pr(C_{gp}^{(1)} = 1, C_{hq}^{(1)} = 1 | F, S, R, \tilde{\Omega})}. \tag{11.23}$$

Fourth, we derive the update rule for Υ. Based on Equations 11.8 and 11.12, we formulate optimizing Υ as the following optimization:

$$\arg\min_{\Upsilon} \sum_{C^{(1)}, C^{(2)}} \sum_{g=1}^{k_1} \sum_{h=1}^{k_2} \sum_{\substack{p:C_{gp}^{(1)}=1, \\ q:C_{hq}^{(2)}=1}} D_{\phi_3}(R_{pq}, \Upsilon_{gh}) \tilde{p}, \tag{11.24}$$

where $\tilde{p}$ denotes $Pr(C_{gp}^{(1)} = 1, C_{hq}^{(2)} = 1 | F, S, R, \tilde{\Omega})$, $1 \leq p \leq n_1$ and $1 \leq q \leq n_2$. Based on Theorem 11.1, we update each Γ_{gh} as follows,

$$\Upsilon_{gh} = \frac{\sum_{p=1}^{n_1} \sum_{q=1}^{n_2} R_{pq} Pr(C_{gp}^{(1)} = 1, C_{hq}^{(2)} = 1 | F, S, R, \tilde{\Omega})}{\sum_{p=1}^{n_1} \sum_{q=1}^{n_2} Pr(C_{gp}^{(1)} = 1, C_{hq}^{(2)} = 1 | F, S, R, \tilde{\Omega})}. \tag{11.25}$$

Algorithm 6 Exponential family MMRC algorithm

Input: A relational data set $\{\{F^{(j)}\}_{j=1}^m, \{S^{(j)}\}_{j=1}^m, \{R^{(ij)}\}_{i,j=1}^m\}$, a set of exponential family distributions (Bregman divergences) assumed for the data set.

Output: Membership Matrices $\{\Lambda^{(j)}\}_{j=1}^m$, attribute expectation matrices $\{\Theta^{(j)}\}_{j=1}^m$, homogeneous relation expectation matrices $\{\Gamma^{(j)}\}_{j=1}^m$, and heterogeneous relation expectation matrices $\{\Upsilon^{(ij)}\}_{i,j=1}^m$.

Method:

1: Initialize the parameters as $\tilde{\Omega} = \{\{\tilde{\Lambda}^{(j)}\}_{j=1}^m, \{\tilde{\Theta}^{(j)}\}_{j=1}^m, \{\tilde{\Gamma}^{(j)}\}_{j=1}^m, \{\tilde{\Upsilon}^{(ij)}\}_{i,j=1}^m\}$.
2: **repeat**
3:    {E-step}
4:    Compute the posterior $Pr(\{C^{(j)}\}_{j=1}^m | \{F^{(j)}\}_{j=1}^m, \{S^{(j)}\}_{j=1}^m, \{R^{(ij)}\}_{i,j=1}^m, \tilde{\Omega})$ using the Gibbs sampler.
5:    {M-step}
6:    **for** $j = 1$ to m **do**
7:       Compute $\Lambda^{(j)}$ using update rule Equation 11.16.
8:       Compute $\Theta^{(j)}$ using update rule Equation 11.21.
9:       Compute $\Gamma^{(j)}$ using update rule Equation 11.23.
10:       **for** $i = 1$ to m **do**
11:          Compute $\Upsilon^{(ij)}$ using update rule Equation 11.25.
12:       **end for**
13:    **end for**
14:    $\tilde{\Omega} = \Omega$
15: **until** convergence

Combining the E-step and M-step, we have a general relational clustering algorithm, Exponential Family MMRC (EF-MMRC) algorithm, which is summarized in Algorithm 6. Since it is straightforward to apply our algorithm derivation to a relational data set of any structure, Algorithm 6 is proposed based on the input of a general relational data set. Despite that the input relational data could have various structures, EF-MMRC works simply as follows: in the E-step, EF-MMRC iteratively updates the posterior probabilities that an object is associated with the clusters (the Markov chain in Section 11.1.2); in the M-step, based on the current cluster association (posterior probabilities), the cluster representatives of attributes and relations are updated as the weighted mean of the observations no matter which exponential distributions are assumed.

Therefore, with the simplicity of the traditional centroid-based clustering algorithms, EF-MMRC is capable of making use of all attribute information and homogeneous and heterogeneous relation information to learn hidden structures from various relational data. Since EF-MMRC simultaneously clusters multi-type interrelated objects, the cluster structures of different types of objects may interact with each other directly or indirectly during the clustering process to automatically deal with the influence propagation. Besides the local cluster structures for each type of objects, the output of EF-MMRC also provides the summary of the global hidden structure for the data, i.e., based on Γ and Υ, we know how the clusters of the same type and different types are related to each other. Furthermore, relational data from different applications may have different probabilistic distributions on the attributes and relations; it is easy for EF-MMRC to adapt to this situation by simply using different Bregman divergences corresponding to different exponential family distributions.

If we assume $O(m)$ types of heterogeneous relations among m types of objects, which is typical in real applications, and let $n = \Theta(n_i)$ and $k = \Theta(k_i)$, the computational complexity of EF-MMRC can be shown to be $O(tmn^2k)$ for t iterations. If we apply the k-means algorithm to each type of nodes individually by transforming the relations into attributes for each type of nodes, the total computational complexity is also $O(tmn^2k)$.

11.1.4 Hard MMRC Algorithm

Due to its simplicity, scalability, and broad applicability, k-means algorithm has become one of the most popular clustering algorithms. Hence, it is desirable to extend k-means to relational data. Some efforts [11, 44, 85, 125] in the literature work focus on this direction. However, these approaches apply to only some special and simple cases of relational data, such as bi-type heterogeneous relational data.

As traditional k-means can be formulated as a hard version of Gaussian mixture model EM algorithm [76], we propose the hard version of MMRC algorithm as a general relational k-means algorithm (from now on, we call Algorithm 6 as soft EF-MMRC), which applies to various relational data.

To derive the hard version MMRC algorithm, we omit soft membership parameters $\Lambda^{(j)}$ in the MMRC model ($C^{(j)}$ in the model provides the hard membership for each object). Next, we change the computation of the posterior probabilities in the E-step to reassignment procedure, i.e., in the E-step, based on the estimation of the current parameters, we reassign cluster labels, $\{C^{(j)}\}_{j=1}^m$, to maximize the objective function in Equation 11.8. In particular, for each object, while fixing the cluster assignments of all other objects, we assign it to each cluster to find the optimal cluster assignment maximizing the objective function in Equation 11.8, which is equivalent to minimizing the Bregman distances between the observations and the corresponding expectation parameters. After all objects are assigned, the reassignment process is repeated until no object changes its cluster assignment between two successive iterations.

In the M-step, we estimate the parameters based on the cluster assignments from the E-step. A simple way to derive the update rules is to follow the derivation in Section 11.1.3 but replace the posterior probabilities with its hard versions. For example, after the E-step, if the object $x_p^{(j)}$ is assigned to the gth cluster, i.e., $C_{gp}^{(j)} = 1$, then the posterior $Pr(C_{gp}^{(1)} = 1|F, S, R, \tilde{\Omega}) = 1$ and $Pr(C_{hp}^{(1)} = 1|F, S, R, \tilde{\Omega}) = 0$ for $h \neq g$.

Using the hard versions of the posterior probabilities, we derive the following update rule for $\Theta^{(j)}$:

$$\Theta_{\cdot g}^{(j)} = \frac{\sum_{p:C_{gp}^{(j)}=1} F_{\cdot p}^{(j)}}{\sum_{p=1}^{n_j} C_{gp}^{(j)}}. \tag{11.26}$$

In the above update rule, since $\sum_{p=1}^{n_1} C_{gp}^{(j)}$ is the size of the gth cluster, $\Theta_{\cdot g}^{(j)}$ is actually updated as the mean of the attribute vectors of the objects assigned to the gth cluster.

Similarly, we have the following update rule for $\Gamma^{(j)}$:

$$\Gamma_{gh}^{(j)} = \frac{\sum_{p:C_{gp}^{(j)}=1,q:C_{hq}^{(j)}=1} S_{pq}^{(j)}}{\sum_{p=1}^{n_j} C_{gp}^{(j)} \sum_{q=1}^{n_j} C_{hq}^{(j)}}, \tag{11.27}$$

i.e., $\Gamma_{gh}^{(j)}$ is updated as the mean of the relations between the objects of the jth type from the gth cluster and the hth cluster.

Each heterogeneous relation expectation parameter $\Upsilon_{gh}^{(ij)}$ is updated as the mean of the objects of the ith type from the gth cluster and of the jth type from the hth cluster,

$$\Upsilon_{gh}^{(ij)} = \frac{\sum_{p:C_{gp}^{(i)}=1,q:C_{hq}^{(j)}=1} R_{pq}^{(ij)}}{\sum_{p=1}^{n_i} C_{gp}^{(i)} \sum_{q=1}^{n_j} C_{hq}^{(j)}}. \tag{11.28}$$

The hard version of EF-MMRC algorithm is summarized in Algorithm 7. It works simply as the classic k-means. However, it is applicable to various relational data under various Bregman distance functions corresponding to various assumptions of probability distributions. Based on the EM framework, its convergence is guaranteed. When applied to some special cases of relational data, it provides simple and new algorithms for some important data mining problems. For example, when applied to the data of one homogeneous relation matrix representing a graph affinity matrix, it provides a simple and new graph partitioning algorithm.

Based on Algorithms 6 and 7, there is another version of EF-MMRC, i.e., we may combine soft and hard EF-MMRC together to have mixed EF-MMRC. For example, we first run hard EF-MMRC several times as initialization, then run soft EF-MMRC.

Algorithm 7 Hard MMRC algorithm

Input: A relational data set $\{\{F^{(j)}\}_{j=1}^m, \{S^{(j)}\}_{j=1}^m, \{R^{(ij)}\}_{i,j=1}^m\}$, a set of exponential family distributions (Bregman divergences) assumed for the data set.

Output: Cluster indicator matrices $\{C^{(j)}\}_{j=1}^m$, attribute expectation matrices $\{\Theta^{(j)}\}_{j=1}^m$, homogeneous relation expectation matrices $\{\Gamma^{(j)}\}_{j=1}^m$, and heterogeneous relation expectation matrices $\{\Upsilon^{(ij)}\}_{i,j=1}^m$.

Method:

1:   Initialize the parameters as
     $\tilde{\Omega} = \{\{\tilde{\Lambda}^{(j)}\}_{j=1}^m, \{\tilde{\Theta}^{(j)}\}_{j=1}^m, \{\tilde{\Gamma}^{(j)}\}_{j=1}^m, \{\tilde{\Upsilon}^{(ij)}\}_{i,j=1}^m\}.$

2:   **repeat**

3:     {E-step}

4:     Based on the current parameters, reassign cluster labels for each objects, i.e., update $\{C^{(j)}\}_{j=1}^m$, to maximize the objective function in Equation 11.8.

5:     {M-step}

6:     **for** $j = 1$ to m **do**

7:       Compute $\Theta^{(j)}$ using update rule Equation 11.26.

8:       Compute $\Gamma^{(j)}$ using update rule Equation 11.27.

9:       **for** $i = 1$ to m **do**

10:        Compute $\Upsilon^{(ij)}$ using update rule Equation 11.28.

11:       **end for**

12:     **end for**

13:     $\tilde{\Omega} = \Omega$

14: **until** convergence

11.2   Spectral Relational Clustering Algorithm

In this section, we derive a spectral clustering algorithm for MTRD under the CFRM model. First, without loss of generality, we redefine the cluster indicator matric $C^{(i)}$ as the following vigorous cluster indicator matrix:

$$C_{pq}^{(i)} = \begin{cases} \frac{1}{|\pi_q^{(i)}|^{\frac{1}{2}}} & \text{if } x_{ip} \in \pi_q^{(i)} \\ 0 & \text{otherwise} \end{cases}$$

where $|\pi_q^{(i)}|$ denotes the number of objects in the qth cluster of $\mathcal{X}^{(i)}$. Clearly $C^{(i)}$ still captures the disjoint cluster memberships and $(C^{(i)})^T C^{(i)} = I_{k_i}$ where I_{k_i} denotes $k_i \times k_i$ identity matrix. Hence our task is the minimization:

$$\min_{\substack{\{(C^{(i)})^T C^{(i)} = I_{k_i}\}_{1 \le i \le m} \\ \{A^{(ij)} \in \mathbb{R}^{k_i \times k_j}\}_{1 \le i < j \le m} \\ \{B^{(i)} \in \mathbb{R}^{k_i \times f_i}\}_{1 \le i \le m}}} L \qquad (11.29)$$

where L is the same as in Equation 5.2.

Then, we prove the following lemma, which is useful in proving our main theorem.

LEMMA 11.1

If $\{C^{(i)}\}_{1\leq i\leq m}$, $\{A^{(ij)}\}_{1\leq i<j\leq m}$, and $\{B^{(i)}\}_{1\leq i\leq m}$ are the optimal solution to Equation 11.29, then

$$A^{(ij)} = (C^{(i)})^T R^{(ij)} C^{(j)} \tag{11.30}$$

$$B^{(i)} = (C^{(i)})^T F^{(i)} \tag{11.31}$$

for $1 \leq i \leq m$.

PROOF The objective function in Equation 11.29 can be expanded as follows:

$$
\begin{aligned}
L = \sum_{1\leq i<j\leq m} & w_a^{(ij)} \mathrm{tr}((R^{(ij)} - C^{(i)} A^{(ij)} (C^{(j)})^T) \\
& (R^{(ij)} - C^{(i)} A^{(ij)} (C^{(j)})^T)^T) + \\
\sum_{1\leq i\leq m} & w_b^{(i)} \mathrm{tr}((F^{(i)} - C^{(i)} B^{(i)})(F^{(i)} - C^{(i)} B^{(i)})^T) \\
= \sum_{1\leq i<j\leq m} & w_a^{(ij)} (\mathrm{tr}(R^{(ij)} (R^{(ij)})^T) + \\
& \mathrm{tr}(A^{(ij)} (A^{(ij)})^T) - 2\mathrm{tr}(C^{(i)} A^{(ij)} (C^{(i)})^T (R^{(ij)})^T)) \\
+ \sum_{1\leq i\leq m} & w_b^{(i)} (\mathrm{tr}(F^{(i)} (F^{(i)})^T) + \mathrm{tr}(B^{(i)} (B^{(i)})^T) \\
& -2\mathrm{tr}(C^{(i)} B^{(i)} (F^{(i)})^T)),
\end{aligned}
\tag{11.32}
$$

where tr denotes the trace of a matrix; the terms $\mathrm{tr}(A^{(ij)}(A^{(ij)})^T)$ and $\mathrm{tr}(B^{(i)}(B^{(i)})^T)$ result from the communicative property of the trace and $(C^{(i)})^T(C^{(i)}) = I_{k_i}$. Based on Equation 11.32, solving $\frac{\partial L}{\partial A^{(ij)}} = 0$ and $\frac{\partial L}{\partial B^{(i)}} = 0$ leads to Equations 11.30 and 11.31. This completes the proof of the lemma. ∎

Lemma 11.1 implies that the objective function in Equation 5.2 can be simplified to the function of only $C^{(i)}$. This leads to the following theorem, which is the basis of our algorithm.

THEOREM 11.2

The minimization problem in Equation 11.29 is equivalent to the following maximization problem:

$$\max_{\substack{\{(C^{(i)})^T C^{(i)} \\ =I_{k_i}\}_{1 \leq i \leq m}}} \sum_{1 \leq i \leq m} w_b^{(i)} \mathrm{tr}((C^{(i)})^T F^{(i)} (F^{(i)})^T C^{(i)}) +$$

$$\sum_{1 \leq i < j \leq m} w_a^{(ij)} \mathrm{tr}((C^{(i)})^T R^{(ij)} C^{(j)} (C^{(j)})^T (R^{(ij)})^T C^{(i)}) \qquad (11.33)$$

PROOF   From Lemma 11.1, we have Equations 11.30 and (11.31). Plugging them into Equation 11.32, we obtain

$$L = \sum_{1 \leq i \leq m} w_b^{(i)} (\mathrm{tr}(F^{(i)} (F^{(i)})^T) -$$

$$\mathrm{tr}((C^{(i)})^T F^{(i)} (F^{(i)})^T C^{(i)})) +$$

$$\sum_{1 \leq i < j \leq m} w_a^{(ij)} (\mathrm{tr}(R^{(ij)} (R^{(ij)})^T) -$$

$$\mathrm{tr}((C^{(i)})^T R^{(ij)} C^{(j)} (C^{(j)})^T (R^{(ij)})^T C^{(i)})). \qquad (11.34)$$

Since in Equation 11.34, $\mathrm{tr}(F^{(i)} (F^{(i)})^T)$ and $\mathrm{tr}(R^{(ij)} (R^{(ij)})^T)$ are constants, the minimization of L in Equation 11.29 is equivalent to the maximization in Equation 11.33. This completes the proof of the theorem.   □

We propose an iterative algorithm to determine the optimal (local) solution to the maximization problem in Theorem 12.3, i.e., at each iterative step we maximize the objective function in Equation 11.33 w.r.t. only one matrix $C^{(p)}$ and fix other $C^{(j)}$ for $j \neq p$ where $1 \leq p, j \leq m$. Based on Equation 11.33, after a little algebraic manipulation, the task at each iterative step is equivalent to the following maximization:

$$\max_{(C^{(p)})^T C^{(p)} = I_{k_p}} \mathrm{tr}((C^{(p)})^T M^{(p)} C^{(p)}), \qquad (11.35)$$

where

$$M^{(p)} = w_b^{(p)} (F^{(p)} (F^{(p)})^T) +$$

$$\sum_{p < j \leq m} w_a^{(pj)} (R^{(pj)} C^{(j)} (C^{(j)})^T (R^{(pj)^T})) +$$

$$\sum_{1 \leq j < p} w_a^{(jp)} ((R^{(jp)})^T C^{(j)} (C^{(j)})^T (R^{(jp)})). \qquad (11.36)$$

Clearly $M^{(p)}$ is a symmetric matrix. Since $C^{(p)}$ is a vigorous cluster indicator matrix, the maximization problem in Equation 11.35 is still NP-hard. However, as in the spectral graph partitioning, if we apply real relaxation to $C^{(p)}$ to let $C^{(p)}$ be an arbitrary orthonormal matrix, it turns out that the maximization in Equation 11.35 has a closed-form solution.

Algorithm 8 Spectral relational clustering

Input: Relation matrices $\{R^{(ij)} \in \mathbb{R}^{n_i \times n_j}\}_{1 \leq i < j \leq m}$, feature matrices $\{F^{(i)} \in \mathbb{R}^{n_i \times f_i}\}_{1 \leq i \leq m}$, numbers of clusters $\{k_i\}_{1 \leq i \leq m}$, weights $\{w_a^{(ij)}, w_b^{(i)} \in R_+\}_{1 \leq i < j \leq m}$.

Output: Cluster indicator matrices $\{C^{(p)}\}_{1 \leq p \leq m}$.

Method:

  1: Initialize $\{C^{(p)}\}_{1 \leq p \leq m}$ with othonormal matrices.
  2: **repeat**
  3:    **for** $p = 1$ to m **do**
  4:       Compute the matrix $M^{(p)}$ as in Equation 11.36.
  5:       Update $C^{(p)}$ by the leading k_p eigenvectors of $M^{(p)}$.
  6:    **end for**
  7: **until** convergence
  8: **for** $p = 1$ to m **do**
  9:    transform $C^{(p)}$ into a cluster indicator matrix by the k-means.
 10: **end for**

THEOREM 11.3

(Ky-Fan thorem) Let M be a symmetric matrix with eigenvalues $\lambda_1 \geq \lambda_2 \geq \ldots \geq \lambda_k$, and the corresponding eigenvectors $U = [u_1, \ldots, u_k]$. Then $\sum_{i=1}^{k} \lambda_i = \max_{X^T X = I_k} \text{tr}(X^T M X)$. Moreover, the optimal X is given by $[u_1, \ldots, u_k]Q$ where Q is an arbitrary orthogonal matrix.

Based on Theorem 12.4 [17], at each iterative step we update $C^{(p)}$ as the leading k_p eigenvectors of the matix $M^{(p)}$. After the iteration procedure converges, since the resulting eigen-matrices are not indicator matrices, we need to transform them into cluster indicator matrices by postprocessing [8,46,138]. In this paper, we simply adopt the k-means for the postprocessing.

The algorithm, called Spectral Relational Clustering (SRC), is summarized in Algorithm 8. By iteratively updating $C^{(p)}$ as the leading k_p eigenvectors of $M^{(p)}$, SRC makes use of the interactions among the hidden structures of different type of objects. After the iteration procedure converges, the hidden structure for each type of objects is embedded in an eigen-matrix. Finally, we postprocess each eigen-matrix to extract the cluster structure.

To illustrate the SRC algorithm, we describe the specific update rules for the tri-type relational data: update $C^{(1)}$ as the leading k_1 eigenvectors of $w_a^{(12)} R^{(12)} C^{(2)} (C^{(2)})^T (R^{(12)})^T$; update $C^{(2)}$ as the leading k_2 eigenvectors of $w_a^{(12)} (R^{(12)})^T C^{(1)} (C^{(1)})^T R^{(12)} + w_a^{(23)} R^{(23)} C^{(3)} (C^{(3)})^T (R^{(23)})^T$; update $C^{(3)}$ as the leading k_3 eigenvectors of $w_a^{(23)} (R^{(23)})^T C^{(2)} (C^{(2)})^T R^{(23)}$.

The computational complexity of SRC can be shown to be $O(tmn^2 k)$ where t denotes the number of iterations, $n = \Theta(n_i)$ and $k = \Theta(k_i)$. For sparse data, it could be reduced to $O(tmzk)$ where z denotes the number of nonzero elements.

The convergence of SRC algorithm can be proved. We describe the main idea as follows. Theorem 12.3 and Equation 11.35 imply that the updates of the matrices in Line 5 of Algorithm 1 increase the objective function in Equation 11.33, and hence equivalently decrease the objective function in Equation 11.29. Since the objective function in Equation 11.29 has the lower bound 0, the convergence of SRC is guaranteed.

11.3   A Unified View to Clustering

In this section, we discuss the connections between existing clustering approaches and the MMRF model and EF-MMRF algorithms. By considering them as special cases or variations of the MMRF model, we show that MMRF provides a unified view to the existing clustering approaches from various important data mining applications.

11.3.1   Semi-Supervised Clustering

Recently, semi-supervised clustering has become a topic of significant interest [14, 124], which seeks to cluster a set of data points with a set of pairwise constraints.

Semi-supervised clustering can be formulated as a special case of relational clustering, clustering on the single-type relational data set consisting of attributes F and homogeneous relations S. For semi-supervised clustering, S_{pq} denotes the pairwise constraint on the pth object and the qth object.

[14] provides a general model for semi-supervised clustering based on Hidden Markov Random Fields (HMRFs). We show that it can be formulated as a special case of the MMRC model. As in [14], we define the homogeneous relation matrix S as follows:

$$S_{pq} = \begin{cases} f_M(x_p, x_q) & \text{if } (x_p, x_q) \in \mathcal{M} \\ f_C(x_p, x_q) & \text{if } (x_p, x_q) \in \mathcal{C} \\ 0 & \text{otherwise} \end{cases}$$

where $\mathcal{M}$ denotes a set of must-link constraints; $\mathcal{C}$ denotes a set of cannot-link constraints; $f_M(x_p, x_q)$ is a function that penalizes the violation of must-link constraint; $f_C(x_p, x_q)$ is a penalty function for cannot-links. If we assume Gibbs distribution [112] for S,

$$Pr(S) = \frac{1}{z_1} \exp(-\sum_{p,q} S_{pq}). \tag{11.37}$$

where z_1 is the normalization constant. Since [14] focuses on only hard clustering, we omit the soft member parameters in the MMRC model to consider hard clustering. Based on Equations 11.37 and 11.3, the likelihood function of hard semi-supervised clustering under MMRC model is

$$L(\Theta|F) = \frac{1}{z}\exp(-\sum_{p,q} S_{pq})\exp(-D_\phi(F, \Lambda C)). \tag{11.38}$$

Since C is an indicator matrix, Equation 11.38 can be formulated as

$$L(\Theta|F) = \frac{1}{z}\exp(-\sum_{p,q} S_{pq})\exp(-\sum_{g=1}^{k}\sum_{p:C_{gp}=1} D_\phi(F_{\cdot p}, \Lambda_{\cdot g})). \tag{11.39}$$

The above likelihood function is equivalent to the objective function of semi-supervised clustering based on HMRFs [14]. Furthermore, when applied to optimizing the objective function in Equation 11.39, hard MMRC provides a family of semi-supervised clustering algorithms similar to HMRF-KMeans in [14]; on the other hand, soft EF-MMRC provides a new and soft version semi-supervised clustering algorithms.

11.3.2 Co-Clustering

Co- or bi-clustering arises in many important applications, such as document clustering, micro-array data clustering. A number of approaches [11, 31, 44, 85] have been proposed for co-clustering. These efforts can be generalized as solving the following matrix approximation problem [86],

$$\arg\min_{C,\Upsilon} \mathfrak{D}(R, (C^{(1)})^T\Upsilon C^{(2)}), \tag{11.40}$$

where $R \in \mathbb{R}^{n_1 \times n_2}$ is the data matrix , $C^{(1)} \in \{0,1\}^{k_1 \times n_1}$ and $C^{(2)} \in \{0,1\}^{k_2 \times n_2}$ are indicator matrices, $\Upsilon \in \mathbb{R}^{k_1 \times k_2}$ is the relation representative matrix, and $\mathfrak{D}$ is a distance function. For example, [44] uses KL-divergences as the distance function; [31, 85] use Euclidean distances.

Co-clustering is equivalent to clustering on relational data of one heterogeneous relation matrix R. Based on Equation 11.8, by omitting the soft membership parameters, maximizing log-likelihood function of hard clustering on a heterogeneous relation matrix under the MMRC model is equivalent to the minimization in Equation 11.40. The algorithms proposed in [11, 31, 44, 85] can be viewed as special cases of hard EF-MMRC. At the same time, soft EF-MMRC provides another family of new algorithms for co-clustering.

Chapter 3 proposes the relation summary network model for clustering k-partite graphs, which can be shown to be equivalent to clustering on relational data of multiple heterogeneous relation matrices. The proposed algorithms in Chapter 9 can also be viewed as special cases of the hard EF-MMRC algorithm.

11.3.3 Graph Clustering

Graph clustering (partitioning) is an important problem in many domains, such as circuit partitioning, VLSI design, and task scheduling. Existing graph partitioning approaches are mainly based on edge cut objectives, such as Kernighan-Lin objective [77], normalized cut [113], ratio cut [28], ratio association [113], and min-max cut [47].

Graph clustering is equivalent to clustering on single-type relational data of one homogeneous relation matrix S. The log-likelihood function of the hard clustering under MMRC model is $-D_\phi(S, (C)^T \Gamma C)$. We propose the following theorem to show that the edge cut objectives are mathematically equivalent to a special case of the MMRC model. Since most graph partitioning objective functions use a weighted indicator matrix such that $CC^T = I_k$, where I_k is an identity matrix, we follow this formulation in the following theorem.

THEOREM 11.4
With restricting Γ to be the form of rI_k for $r > 0$, maximizing the log-likelihood of hard MMRC clustering on S under normal distribution, i.e.,

$$\max_{C \in \{0,1\}^{k \times n}, CC^T = I_k} -\|S - (C)^T (rI_k)C\|^2, \tag{11.41}$$

is equivalent to the trace maximization

$$\max \operatorname{tr}(CSC^T), \tag{11.42}$$

where tr denotes the trace of a matrix.

PROOF	Let L denote the objective function in Equation 11.41.

$$
\begin{aligned}
L &= -\|S - rC^T C\|^2 \\
&= -\operatorname{tr}((S - rC^T C)^T (S - rC^T C)) \\
&= -\operatorname{tr}(S^T S) + 2r\operatorname{tr}(C^T CS) - r^2 \operatorname{tr}(C^T CC^T C) \\
&= -\operatorname{tr}(S^T S) + 2r\operatorname{tr}(CSC^T) - r^2 k.
\end{aligned}
$$

The above deduction uses the property of trace $\operatorname{tr}(XY) = \operatorname{tr}(YX)$. Since $\operatorname{tr}(S^T S)$, r, and k are constants, the maximization of L is equivalent to that of $\operatorname{tr}(CSC^T)$. The proof is completed.					☐

Since it is shown in the literature [40] that the edge cut objectives can be formulated as the trace maximization, Theorem 11.4 states that the edge-cut-based graph clustering is equivalent to MMRC model under normal distribution with the diagonal constraint on the parameter matrix Γ. This connection provides not only a new understanding for graph partitioning but also a family of new algorithms (soft and hard MMRC algorithms) for graph clustering.

Finally, we point out that the MMRC model does not exclude traditional attribute-based clustering. When applied to an attribute data matrix under Euclidean distances, hard MMRC algorithm is actually reduced to the classic k-means; soft MMRC algorithm is very close to the traditional mixture model EM clustering except that it does not involve mixing proportions in the computation.

In summary, the MMRC model provides a principal framework to unify various important clustering tasks, including traditional attributes-based clustering, semi-supervised clustering, co-clustering, and graph clustering; soft and hard EF-MMRC algorithms unify a number of state-of-the-art clustering algorithms and at the same time provide new solutions to various clustering tasks.

Chapter 12

Multiple-View Relational Data Clustering

In Chapter 6, we proposed a general model for multiple-view unsupervised learning. The proposed model introduces the concept of mapping function to make the different patterns from different pattern spaces comparable and hence an optimal pattern can be learned from the multiple patterns of multiple representations. In this chapter, under the model we formulate two more specific models for two important cases of unsupervised learning: clustering and spectral dimensionality reduction; we derive an iterating algorithm for multiple-view clustering, and a simple algorithm providing a global optimum to multiple spectral dimensionality reduction. We also extend the proposed model and algorithms to evolutionary clustering and unsupervised learning with side information.

12.1   Algorithm Derivation

In this section, we derive algorithms for multiple-view clustering and multiple-view spectral embedding. To avoid the clutter, in the following algorithm derivations we omit the weights w_i. However, all the derivation can be extended to weighted cases. Based on Definitions 6.2 and 6.3, Our task is to solve the following two optimizations: multiple-view clustering,

$$\min_{\substack{B \geq 0, B\mathbf{1}=\mathbf{1} \\ \{P^{(i)} \geq 0\}_{i=1}^m}} \sum_{i=1}^m GI(A^{(i)} \| BP^{(i)}), \tag{12.1}$$

and multiple-view spectral embedding,

$$\min_{\substack{B^T B = I, \\ \{P^{(i)}\}_{i=1}^m}} \sum_{i=1}^m \|A^{(i)} - BP^{(i)}\|^2, \tag{12.2}$$

where $B \geq 0$ and $P^{(i)} \geq 0$ denote that B and $P^{(i)}$ are nonnegative.

123

12.1.1   Multiple-View Clustering Algorithm

Since Generalized I-divergence is a separable distance function, by letting $A = [A^{(1)}, \ldots, A^{(m)}]$ and $P = [P^{(1)}, \ldots, P^{(m)}]$, we write the objective function in Equation 12.1 in a pure matrix form as follows:

$$\min_{\substack{B \geq 0, B1 = 1 \\ P \geq 0}} GI(A \| BP). \qquad (12.3)$$

Note that $A \in \mathbb{R}^{n \times r}$ and $P \in \mathbb{R}^{k \times r}$ with $r = \sum_{i=1}^{m} k_i$.

To the best of our knowledge, there is no efficient way to find the global optimum to the constrained non-convex optimization in Equation 12.3. We derive an alternating algorithm for Equation 12.3, which iteratively updates B and P until it converges to a local optimum.

First, we fix P to update B. To deal with the constraint $B1 = 1$ efficiently, we transform it to a "soft" constraint by adding a penalty term, $\alpha GI(1 \| B1)$, to the objective function, where α is a positive constant. Therefore, we obtain the following optimization:

$$\min_{B \geq 0} GI(A \| BP) + \alpha GI(1 \| B1). \qquad (12.4)$$

We derive an efficient updating rule for B based on the bound optimization procedure [36,107]. The basic idea is to construct an auxiliary function which is a convex upper bound for the original objective function based on the solution obtained from the previous iteration. Then, a new solution to the current iteration is obtained by minimizing this upper bound. The definition of an auxiliary function is given as follows.

DEFINITION 12.1 $h(S, S^t)$ *is an auxiliary function for* $f(S)$ *if* $h(S, S^t) \geq f(S)$ *and* $h(S, S) = f(S)$.

The auxiliary function is useful due to the following lemma.

LEMMA 12.1
If h is an auxiliary function, then f is nonincreasing under the updating rule,

$$S^{t+1} = \arg\min_{S} h(S, S^t). \qquad (12.5)$$

PROOF $f(S^{t+1}) \leq h(S^{t+1}, S^t) \leq h(S^t, S^t) \leq f(S^t).$ □

The key step of the derivation is to design auxiliary functions. We propose an auxiliary function for B in the following lemma.

LEMMA 12.2

$$h(B, \tilde{B}) = \sum_{ij}(\sum_g (B_{ig}P_{gj}) - A_{ij}\sum_g \frac{\tilde{B}_{ig}P_{gj}}{[\tilde{B}P]_{ij}}\log B_{ig})$$

$$+ \alpha \sum_i ([B1]_i - \sum_g (\frac{\tilde{B}_{ig}}{[\tilde{B}1]_i}\log B_{ig})) + t(\tilde{B})$$

is an auxiliary function for

$$f(B) = GI(A\|BP) + \alpha GI(\mathbf{1}\|B1), \qquad (12.6)$$

where $t(\tilde{B})$ is a function w.r.t. $\tilde{B}$ s.t.

$$t(\tilde{B}) = \sum_{ij}(A_{ij}\log A_{ij} - A_{ij} - A_{ij}\sum_g (\frac{\tilde{B}_{ig}P_{gj}}{[\tilde{B}P]_{ij}}\log \frac{[\tilde{B}P]_{ij}}{\tilde{B}_{ig}}))$$

$$- n\alpha - \alpha \sum_i \sum_g (\frac{\tilde{B}_{ig}}{[\tilde{B}1]_i}\log(\frac{[\tilde{B}1]_i}{\tilde{B}_{ig}})).$$

PROOF

$$f(B) = \sum_{ij}(A_{ij}\log\frac{A_{ij}}{[BP]_{ij}} - A_{ij} + [BP]_{ij})$$

$$+ \alpha\sum_i(-\log[B1]_i - 1 + [B1]_i)$$

$$\leq \sum_{ij}(A_{ij}\log A_{ij} - A_{ij} + \sum_g(B_{ig}P_{gj})$$

$$- A_{ij}\sum_g(\frac{\tilde{B}_{ig}P_{gj}}{[\tilde{B}P]_{ij}}\log(\frac{[\tilde{B}P]_{ij}}{\tilde{B}_{ig}P_{gj}}B_{ig}P_{gj})))$$

$$+ \alpha\sum_i([B1]_i - 1 - \sum_g(\frac{\tilde{B}_{ig}}{[\tilde{B}1]_i}\log(\frac{[\tilde{B}1]_i}{\tilde{B}_{ig}}B_{ig})))$$

$$= h(B, \tilde{B}).$$

During the above deduction, we use Jensen's inequality and concavity of the log function. □

The following theorem provides the updating rule for B.

THEOREM 12.1
The objective function $f(B)$ in Equation 12.6 is nonincreasing under the

updating rule,

$$B_{ig} = \tilde{B}_{ig} \frac{\sum_j (A_{ij} \frac{P_{gj}}{[\tilde{B}P]_{ij}}) + \frac{\alpha}{[\tilde{B}1]_i}}{\sum_j P_{gj} + \alpha} \tag{12.7}$$

where $\tilde{B}$ denotes the solution from the previous iteration.

PROOF   Based on Lemma 12.2, take the derivative of $h(B, \tilde{B})$ w.r.t. B_{ig} to obtain

$$\frac{\partial h(B, \tilde{B})}{\partial B_{ig}} = \sum_j (P_{gj} - A_{ij} \frac{P_{gj} \tilde{B}_{ig}}{[\tilde{B}P]_{ij} B_{ig}})$$

$$+ \alpha - \alpha \frac{\tilde{B}_{ig}}{[\tilde{B}1]_i B_{ig}}.$$

Solve $\frac{\partial h(B,\tilde{B})}{\partial B_{ig}} = 0$ to obtain Equation 12.7. By Lemma 12.1, the proof is completed. □

Similarly, we present the following theorems to derive the updating rule for P.

LEMMA 12.3

$$h(P, \tilde{P}) = \sum_{ij} (A_{ij} \log A_{ij} - A_{ij} + \sum_g (B_{ig} P_{gj}))$$

$$- A_{ij} \sum_g (\frac{B_{ig} \tilde{P}_{gj}}{[B\tilde{P}]_{ij}} \log \frac{[B\tilde{P}]_{ij} P_{gj}}{\tilde{P}_{gj}}))$$

is an auxiliary function for

$$f(P) = GI(A||BP). \tag{12.8}$$

THEOREM 12.2
The objective function $f(P)$ in Equation 12.8 is nonincreasing under the updating rule

$$P_{gj} = \tilde{P}_{gj} \frac{\sum_i \frac{A_{ij} B_{ig}}{[B\tilde{P}]_{ij}}}{\sum_i B_{ig}}. \tag{12.9}$$

Following the way to prove Lemma 12.2 and Theorem 12.1, it is not difficult to prove the above theorems. We omit details here.

We call the algorithm as the Multiple-view clustering (MVC) algorithm, which is summarized in Algorithm 9. The complexity of the algorithm is

---

Algorithm 9 Multiple-view clustering

Input: A set of clustering pattern matrices denoted as $A = [A^{(1)} \ldots A^{(m)}]$ and a positive integer k.

Output: A clustering pattern matrix B and a set of mapping matrices $P = [P^{(1)} \ldots P^{(2)}]$.

Method:

1: Initialize B and P.

2: **repeat**

3:

$$B_{ig} = B_{ig} \frac{\sum_j (A_{ij} \frac{P_{gj}}{[BP]_{ij}}) + \frac{\alpha}{[B1]_i}}{\sum_j P_{gj} + \alpha}.$$

4:

$$P_{gj} = \tilde{P}_{gj} \frac{\sum_i \frac{A_{ij} B_{ig}}{[B\tilde{P}]_{ij}}}{\sum_i B_{ig}}$$

5: **until** convergence

$O(tnrk)$ for t iterations and it can be further reduced for sparse data. The convergence of the MVC algorithm is guaranteed by Theorems 12.1 and 12.2.

12.1.2 Multiple-View Spectral Embedding Algorithm

Since Frobenius norm is also a separable distance function, by letting $A = [A^{(1)}, \ldots, A^{(m)}]$ and $P = [P^{(1)}, \ldots, P^{(m)}]$, we can rewrite the objective function of multiple-view spectral embedding as follows:

$$\min_{B^T B = I} ||A - BP||^2. \tag{12.10}$$

We show how to derive an algorithm which provides the global optimum to the optimization in Equation 12.10. First, we use KKT condition to derive the following lemma.

LEMMA 12.4
If B and P are the optimal solution to Equation 12.10, then

$$P = B^T A \tag{12.11}$$

PROOF Let $f(B, P)$ denote the objective function in Equation 12.10. $f(B, P)$ can be expanded as follows:

$$f(B, P) = tr((A - BP)^T (A - BP))$$
$$= tr(A^T A) - 2tr(P^T B^T A) + tr(P^T P).$$

In the last step of the above deduction, we use $B^T B = I$. Taking the derivative of $f(B, P)$ w.r.t. P, we obtain

$$\frac{\partial f(B, P)}{\partial P} = -2B^T A + 2P. \qquad (12.12)$$

According to KKT condition, we solve $\frac{\partial f(B,P)}{\partial P} = 0$ to obtain Equation 12.11. The proof is completed. ☐

Based on Lemma 12.4, if we know the optimal B, we can easily obtain the optimal P by Equation 12.11. Based on the following theorem, we derive the closed form for the optimal B.

THEOREM 12.3
The minimization problem in Equation 12.10 is equivalent to the following maximization problem:

$$\max_{B^T B = I} tr(B^T A A^T B). \qquad (12.13)$$

PROOF Based on Lemma 12.4, we substitute Equation 12.11 into the objective function in 12.10 to obtain

$$f(B, P) = tr(A^T A) - tr(B^T A A^T B).$$

Since A is a constant matrix, minimization in Equation 12.10 is equivalent to the maximization in Equation 12.13. This completes the proof of the theorem. ☐

It turns out that the maximization in Equation 12.13 has a closed-form solution.

THEOREM 12.4
(Ky-Fan thorem [17]) Let M be a symmetric matrix with eigenvalues $\lambda_1 \geq \lambda_2 \geq \ldots \geq \lambda_k$, and the corresponding eigenvectors $U = [\mathbf{u}_1, \ldots, \mathbf{u}_k]$. Then $\sum_{i=1}^{k} \lambda_i = \max_{X^T X = I_k} tr(X^T M X)$. Moreover, the optimal X is given by $[\mathbf{u}_1, \ldots, \mathbf{u}_k]Q$ where Q is an arbitrary orthogonal matrix .

Based on Theorems 12.3 and 12.4, the optimal B is given as the leading k eigenvectors of the matrix AA^T.

Hence, we have a simple algorithm for Multiple-View Spectral Embedding (MVSE), which is summarized in Algorithm 10. The derivation procedure of Algorithm 10 can be extended to deriving algorithms for other types of multiple-view dimensionally reduction, such as multiple-view sparse coding [84].

Algorithm 10 Multiple-view spectral embedding

Input: A set of spectral embedding matrices denoted as $A = [A^{(1)}, \ldots, A^{(m)}]$ and a positive integer k.

Output: A spectral embedding matrix B and a set of mapping matrices $P = [P^{(1)}, \ldots, P^{(2)}]$.

Method:

1: $B = [\mathbf{u}_1 \ldots \mathbf{u}_k]$ where $\{\mathbf{u}_i\}_{i=1}^k$ are the leading k eigenvectors of the matrix AA^T.

2: $P = B^T A$

12.2 Extensions and Discussions

In this section, we discuss how to extend the proposed model and algorithms to two important related fields: unsupervised learning with side information and evolutionary clustering.

12.2.1 Evolutionary Clustering

In the literature of multiple-view learning, there is a popular assumption that multiple representations are independent [19, 37, 103], though in real applications this assumption may not always hold true. On the other hand, our model and algorithms do not make this assumption. Hence, they are applicable to a wider range of data. Furthermore, this makes it easy to extend our model and algorithms to a fairly new field, evolutionary unsupervised learning.

Evolutionary clustering arises in the dynamic application scenarios, where the objects to be clustered evolve with time. In evolutionary clustering, we need to address two issues at each time step: the current clustering pattern should depend mainly on the current data features; on the other hand, the current clustering pattern should not deviate dramatically from the most recent history, i.e., we expect a certain level of temporal smoothness between clusters in successive time steps [30].

If we treat data features at each time step as a new representation for the data objects, evolutionary data can be formulated as a special case of multiple-view data. Clearly, representations are not independent here. Furthermore, our task in evolutionary clustering is to learn a series of clustering patterns with temporal smoothness, instead of one clustering pattern based on all historic data. Our multiple-view clustering model and algorithm can be easily extended to the evolutionary clustering case. We describe the procedure as follows:

- At each time step t, the basic clustering pattern $A^{(t)}$ is directly learned from the data features at time t by a specified clustering algorithm.

- A temporal smooth clustering pattern $B^{(t)}$ is given by calling Algorithm 9 s.t.
  $\text{MVC}([B^{(t-1)}, A^{(t)}], k)$, where k is the desired number of clusters for $B^{(t)}$.

The above procedure outputs a series of clustering patterns $\{B^{(t)}\}_{t=1}^{m}$, which depend on the current data features and also consistent with the most recent clustering pattern. This procedure also allows the number of clusters to change with time. Another interesting property about it is that unlike existing efforts focusing on specific type of clustering algorithms, such as evolutionary k-means [27] and evolution spectral clustering [30], it is flexible to adopt any clustering algorithms. The proposed procedure can also be extended to other types of evolutionary unsupervised learning, which have not been touched in the literature.

12.2.2   Unsupervised Learning with Side Information

Unsupervised learning with side information, such as semi-supervised clustering [14, 124], has become a topic of significant interest, which seeks to do unsupervised clustering by incorporating background knowledge, such as partial labeled data or pairwise constraints for data points.

Existing literature on unsupervised clustering with side information focuses on how to make use of side information inside a certain type of clustering algorithm. We discuss how our model and algorithm provide a different style of a solution. The basic idea is that we treat side information as patterns, which is provided by domain experts instead of learned from data. Specifically, we formulate the side information as pattern matrices and use them with pattern matrices from multiple representations together as input to our algorithms. Then, the side information is incorporated into the final optimal pattern.

We describe an example on how to formulate the side information as pattern matrices. Assume that we have partially labeled data. We can represent this side information by assigning the labeled data objects to the clusters with a certain probability and assigning the average probability to the unlabeled data points. In the following side information pattern matrix,

Example 3

$$
A^{(1)} = \begin{bmatrix} 1 & 0 & 0 \\ 0.8 & 0.1 & 0.1 \\ 1/3 & 1/3 & 1/3 \\ 1/3 & 1/3 & 1/3 \\ 1/3 & 1/3 & 1/3 \\ 1/3 & 1/3 & 1/3 \end{bmatrix},
$$

we have two labeled data objects such that the first one is labeled as cluster 1 with probability 1 and the second one is labeled as cluster 1 with probability

0.8; the other data objects are unlabeled. For the pairwise constraint information, they can also be formulated as pattern matrix as follows: assigning the data objects with must-links into the same cluster, the data objects with cannot-links into different clusters, and unconstrained data objects into each cluster with the average probability.

Chapter 13

Evolutionary Data Clustering

In Chapter 7, we proposed three models for evolutionary clustering, DPChain, HDP-EVO, and HDP-HTM. In this chapter, based on Markov Chain Monte Carlo (MCMC) method, we present the algorithms to estimate parameters for those models.

13.1 DPChain Inference

Given the DPChain model, we use Markov Chain Monte Carlo (MCMC) method [92] to sample the cluster assignment $z_{t,i}$ for each data item at time t. Specifically, following Gibbs sampling [26], the aim is to sample the posterior cluster assignment $z_{t,i}$, given the whole data collection $\mathbf{x}_t$ at time t, the history assignment $\{\mathbf{z}_1 \ldots \mathbf{z}_{t-1}\}$, and other assignment $\mathbf{z}_{t,-i}$ at the current time.

We denote $\mathbf{x}_{t,-i}$ as all the data at time t except for $x_{t,i}$. The posterior of the cluster assignment is determined by Bayes rule:

$$p(z_{t,i} = k | \mathbf{x}_t, \mathbf{z}_{t,-i}, \mathbf{z}_1, \ldots \mathbf{z}_{t-1}) \propto \\ p(x_{t,i} | \mathbf{z}_{t,-i}, \mathbf{z}_1, \ldots \mathbf{z}_{t-1}, \mathbf{x}_k^{-i}) p(z_{t,i} = k | \mathbf{z}_1, \ldots \mathbf{z}_{t-1}, \mathbf{z}_{t,-i}), \quad (13.1)$$

where $\mathbf{x}_k^{-i} = \{x_{t,j} : z_{t,j} = k, j \neq i\}$ donates all the data at time t assigned to cluster k except for $x_{t,i}$.

Since $z_{t,i}$ is conditionally indendent of $\mathbf{x}_{t,-i}$ given all the history assignment and the current time assignment except for $x_{t,i}$, we omit $\mathbf{x}_{t,-i}$ at the second term of the right-hand side of Equation 13.1. Further, denote $f_k^{-i}(x_{t,i})$ as the first term of the right-hand side of Equation 13.1, which is the conditional likelihood of $x_{t,i}$ on cluster k, given the other data associated with k and other cluster assignment.

If k is an existing cluster,

$$f_k^{-i}(x_{t,i}) = \int f(x_{t,i} | \phi_{t,k}) \cdot h(\phi_{t,k} | \{x_{t,j} : z_{t,j} = k, j \neq i\}) d\phi_{t,k}, \quad (13.2)$$

where $h(\phi_{t,k} | \{x_j : z_{t,j} = k, j \neq i\})$ is the posterior distribution of parameter $\phi_{t,k}$ given observation $\{x_{t,j} : z_{t,j} = k, j \neq i\}$. If F is conjugate to H, the posterior of $\phi_{t,k}$ is still in the distribution family of H. Then we can integrate out $\phi_{t,k}$ to compute $f_k^{-i}(x_{t,i})$. Here we only consider the conjugate case because our experiments reported in this paper are based on this case. For the nonconjugate case, a similar inference method may be obtained [93].

For a new cluster k, it is equivalent to compute the marginal likelihood of $x_{t,i}$ by integrating out all the parameters sampled from H.

$$f_k^{-i}(x_{t,i}) = \int f(x_{t,i}|\phi_{t,k})dH(\phi_{t,k}). \tag{13.3}$$

Finally, the posterior cluster assignment in the conjugate case is given as

$$p(z_{t,i} = k|\mathbf{x}_t, \mathbf{z}_{t,-i}, \mathbf{z}_1, \dots \mathbf{z}_{t-1}) \propto$$
$$\begin{cases} \frac{w_{t,k}+n_{t,k}^{-i}}{\alpha+\sum_{j=1}^{K_t} w_{t,j}+n_t-1} f_k^{-i}(x_{t,i}) & \text{if } k \text{ is an existing cluster} \\ \frac{\alpha}{\alpha+\sum_{j=1}^{K_t} w_{t,j}+n_t-1} f_k^{-i}(x_{t,i}) & \text{if } k \text{ is a new cluster.} \end{cases} \tag{13.4}$$

13.2 HDP-EVO Inference

Again we use Gibbs sampling [26] for the two-level CRP for HDP-EVO inference. Figure 13.1 illustrate how the clusters at the global and the local levels correspond to each other. First, we specify how to assign $x_{t,i}$ (which may be considered as the last data item by the exchangability) to tab:

$$p(tab_{t,i}|\mathbf{x}_t, tab_{t,1}, \dots, tab_{t,i-1}, \mathbf{K}) \propto$$
$$p(tab_{t,i}|tab_{t,1}, \dots, tab_{t,i-1})p(x_{t,i}|x_{t,-i}, tab_{t,1}, \dots, tab_{t,i-1}, \mathbf{K}). \tag{13.5}$$

For the second-level CRP, We denote the conditional likelihood $f_{k_{t,tab}}^{-i}(x_{t,i})$ as the second term of the right hand side of Equation 13.5.

For an existing table tab which belongs to global cluster $k_{t,tab}$, the conditional likelihood of $x_{t,i}$ given other data under cluster $k_{t,tab}$ indexed from $tab, f_{k_{t,tab}}^{-i}(x_{t,i})$ is the same as that is Equation 13.2 with cluster k replaced with $k_{t,tab}$.

For a new table tab, we first sample the table from the global cluster, the conditional likelihood of $x_{t,i}$ under the cluster k becomes:

$$f_{k_{t,tab}}^{-i}(x_{t,i}) =$$
$$\begin{cases} \frac{m_{t,k}^{-tab}+w_{t,k}}{\gamma+m_t-1+\sum_{j=1}^{K_t} w_{t,j}} f_k^{-i}(x_{t,i}) & k \text{ is an existing cluster} \\ \frac{\gamma}{\gamma+m_t-1+\sum_{j=1}^{K_t} w_{t,j}} f_k^{-i}(x_{t,i}) & k \text{ is a new cluster,} \end{cases} \tag{13.6}$$

where $f_k^{-i}(x_{t,i})$ under new cluster k is the marginal likelihood for a new global cluster k from Equation 13.3 with $\phi_{t,k}$ replaced with ϕ_k.

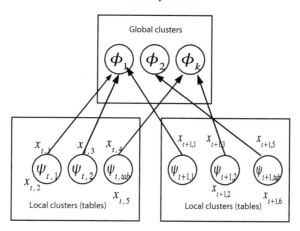

FIGURE 13.1: The illustrated example of global and local cluster correspondence

Finally, we sample $x_{t,i}$ from table tab as follows:

$$
p(tab_{t,i}|\mathbf{x}_t, tab_{t,1}, \ldots, tab_{t,i-1}, \mathbf{K}) \propto
$$
$$
\begin{cases}
\frac{n_{t,tab}^{-i}}{\alpha+n_t-1} f_{k_{t,tab}}^{-i}(x_{t,i}) & \text{if } tab \text{ is an existing table} \\
\frac{\alpha}{\alpha+n_t-1} f_{k_{t,tab}}^{-i}(x_{t,i}) & \text{if } tab \text{ is a new table.}
\end{cases}
\tag{13.7}
$$

Similarly, to sample a table tab from a global cluster k, we have

$$
\begin{aligned}
& p(k_{t,tab}|\mathbf{x}_t, tab_{t,1}, \ldots, tab_{t,i}, \mathbf{K} \setminus k_{t,tab}) \propto \\
& p(k_{t,tab}|\mathbf{K} \setminus k_{t,tab}) p(\mathbf{x}_{t,tab}|\mathbf{x}_{t,-tab}, k_{t,tab}, \mathbf{K} \setminus k_{t,tab}),
\end{aligned}
\tag{13.8}
$$

where $\mathbf{x}_{t,tab}$ denotes all the data belonging to table tab at time t, and $\mathbf{x}_{t,-tab} = \mathbf{x}_t \setminus \mathbf{x}_{t,tab}$ denotes the remaining data except those in table tab.

We denote the second term of the right-hand side of Equation 13.8 as $f_k^{-tab}(\mathbf{x}_{t,tab})$, which means the conditional likelihood of all the data in table tab, given other tables' data at time t, under cluster k.

For an existing global and new cluster k, we have the likelihood

$$
f_k^{-tab}(\mathbf{x}_{t,tab}) = \prod_{i:x_{t,i}\in tab} f_k^{-i}(x_{t,i}).
\tag{13.9}
$$

Finally, we assign a table tab to a global cluster k as follows:

$$
p(k_{t,tab}|\mathbf{x}_t, \mathbf{tab}_t, \mathbf{K} \setminus k_{t,tab}) \propto
$$
$$
\begin{cases}
\frac{m_{t,k}^{-tab}+w_{t,k}}{\gamma+m_t-1+\sum_{j=1}^{K_t} w_{t,j}} f_k^{-tab}(\mathbf{x}_{t,tab}) & \text{if } k \text{ is an existing cluster} \\
\frac{\gamma}{\gamma+m_t-1+\sum_{j=1}^{K_t} w_{t,j}} f_k^{-tab}(\mathbf{x}_{t,tab}) & \text{if } k \text{ is a new cluster,}
\end{cases}
\tag{13.10}
$$

where $\mathbf{tab}_t$ is the set of all the tables at time t.

For both DPChain and HDP-EVO models, there are hyperparameters that must be estimated. We use the EM method [39] to learn these parameters. Specifically, for DPChain, the hyperparameters are (α, η). According to Equation 7.3, updating η results directly in updating $w_{t,k}$. Consequently, we actually update the hyperparameters $\Theta = (\alpha, w_{t,k})$. Following [52], α is sampled from the Gamma distribution at each iteration in the Gibbs sampling in the E-step. In the M-step, similar to [140], we update $w_{t,k}$ by maximizing the cluster assignment likelihood. Suppose that, at an iteration, there are K clusters.

$$w_{t,k}^{new} = \frac{n_{t,k}}{\alpha + n_t - 1} \cdot \sum_{j=1}^{K} w_{t,j}^{old}. \tag{13.11}$$

Thus, the EM framework is as follows:

- At time t, initialize parameters Θ and $z_{t,i}$.

- E-step: Sample α from Gamma Distribution. Sample cluster assignment $z_{t,i}$ for data item $x_{t,i}$ by Equation 13.4.

- M-Step: Update $w_{t,k}$ by (13.11).

- Iterate the E-step and the M-step until the EM converges.

For HDP-EVO, the hyperparameters are $\Theta = (\alpha, \gamma, \lambda)$, Similar parameter learning may be obtained using an EM again.

13.3 HDP-HTM Inference

We denote $n_{i,j}^t$ as the number of the state transitions from state i to j between two adjacent times up to time t. Let $n_{t,k}$ be the number of the data items belonging to cluster k at time t, $n_{t,k}^{-i}$ be the number of the data items belonging to cluster k except $x_{t,i}$ at time t, and n_t be the number of all the data items at time t. Similar to HDP [121], let $m_{t,k}$ be the number of the tables (i.e., the local clusters) belonging to the global cluster k at time t, and m_k be the number of the tables (i.e., the local clusters) belonging to the global cluster k across all the times. Finally, let x_t be the data collection at time t.

In order to handle an infinite or arbitrary number of the states (i.e., clusters), we adopt the stick-breaking mechanism similar to what we have done in Section 7.5.3. Assume that there are K existing clusters. The global mixture proportion $\beta = \{\beta_1, \ldots, \beta_K, \beta_u\}$ with β_u being the proportion for an unrepresented cluster; when a new cluster is instantiated, the vector β is updated according to the stick-breaking construction in Equation 7.12 to ensure the normalized summation equal to 1 with the probability 1. In addition, the transition probability matrix Π is in the dimension of $K+1$ by $K+1$, resulting in

$\boldsymbol{\omega}_t$ also in dimension of 1 by $K+1$ with the last element $\omega_{t,u}$ as the proportion of the unrepresented cluster.

Here, we adopt the EM [39] framework to learn the model by combining Markov Chain Monte Carlo (MCMC) method [92] to make an inference for the auxiliary variable $\boldsymbol{m}$ and other necessary variables at the E-step and maximum likelihood estimation of $\boldsymbol{\Pi}$ at the M-step. Similar to HDP with the direct assignment posterior sampling, we also need to sample the global mixture proportion $\boldsymbol{\beta}$ from $\boldsymbol{m}$. We update the transition probability matrix $\boldsymbol{\Pi}$ by the counter statistic $n_{i,j}^t$ up to time t according to Equation 7.13. We no longer need to sample $\boldsymbol{\theta}_t$ because we may just sample the cluster assignment $\boldsymbol{z}_t$ at time t by integrating out $\boldsymbol{\theta}$. Similarly, by the conjugacy of h and f, it is not necessary to sample the parameter ϕ_k for cluster k.

Sampling $\boldsymbol{z}_t$

Since $\boldsymbol{\theta}_t$ is distributed as $DP(\alpha, \boldsymbol{\omega}_t)$, while the cluster indicator $z_{t,i}$ is in a multinomial distribution with the parameter $\boldsymbol{\theta}_t$, it is convenient to integrate out $\boldsymbol{\theta}_t$ by the conjugacy property. Thus, the conditional probability of the current cluster assignment $z_{t,i}$ for the current data item $x_{t,i}$ given the other assignments $\boldsymbol{z}_{t,-i} = \boldsymbol{z}_t \setminus z_{t,i}$ and the Dirichlet parameters $\boldsymbol{\omega}_t$ and α is

$$p(z_{t,i} = k | \boldsymbol{z}_{t,-i}, \boldsymbol{\omega}_t, \alpha) = \begin{cases} \frac{n_{t,k}^{-i} + \alpha \omega_{t,k}}{n_t - 1 + \alpha} & \text{if } k \text{ is an existing cluster.} \\ \frac{\alpha \omega_{t,u}}{n_t - 1 + \alpha} & \text{if } k \text{ is a new cluster} \end{cases} \quad (13.12)$$

By Gibbs sampling [26], we need to compute the conditional probability $z_{t,i}$ given the other cluster assignment $\boldsymbol{z}_{t,-i}$, the observation $\boldsymbol{x}_t$ at time t, and the Dirichlet parameters $\boldsymbol{\omega}_t$ and α.

$$p(z_{t,i} = k | \boldsymbol{z}_{t,-i}, \boldsymbol{x}_t, \boldsymbol{\omega}_t, \alpha) = \\ \begin{cases} \frac{n_{t,k}^{-i} + \alpha \omega_{t,k}}{n_t - 1 + \alpha} f_k^{-i}(x_{t,i}) & \text{if } k \text{ is an existing cluster,} \\ \frac{\alpha \omega_{t,u}}{n_t - 1 + \alpha} f_k^{-i}(x_{t,i}) & \text{if } k \text{ is a new cluster} \end{cases} \quad (13.13)$$

where $f_k^{-i}(x_{t,i})$ is the conditional likelihood of $x_{t,i}$ given the other data items $\boldsymbol{x}_{t,-i}$ under cluster k, which by the conjugacy property of h and f could be computed by integrating out the cluster parameter ϕ_k for cluster k.

$$f_k^{-i}(x_{t,i}) = \int f(x_{t,i} | \phi_k) \cdot h(\phi_k | \{x_{t,j} : z_{t,j} = k, j \neq i\}) d\phi_k. \quad (13.14)$$

Sampling $\boldsymbol{m}$

Again similar to HDP, in order to sample $\boldsymbol{m}$, we must first sample $\boldsymbol{m}_t$, the number of the tables (i.e., the local clusters) for the clusters at time t [121]. After sampling of $\boldsymbol{z}_t$, $n_{t,k}$ is updated accordingly. By [7, 121], we may sample $\boldsymbol{m}$ according to the following Gibbs sampling [26]:

$$p(m_{t,k} = m | \boldsymbol{z}_t, \boldsymbol{m}^{-t,k}, \boldsymbol{\beta}, \alpha) = \frac{\Gamma(\alpha \beta_k)}{\Gamma(\alpha \beta_k + n_{t,k})} S(n_{t,k}, m)(\alpha \beta_k)^m, \quad (13.15)$$

where $\boldsymbol{m}^{-t,k} = \boldsymbol{m} \setminus m_{t,k}$

Sampling β

Given $\boldsymbol{m}$, the posterior distribution of β is

$$\beta | \boldsymbol{m}, \gamma \sim Dir(m_1, \dots, m_K, \gamma), \tag{13.16}$$

where K is the number of the existing clusters up to time t. Consequently, it is trivial to sample β according to Equation 13.16.

Updating the Transition Matrix Π

After we have the knowledge of the sequence of the states and observations at different times, especially the new knowledge at time t, we may adopt the maximum likelihood estimation to construct the posterior transition probability matrix Π. After sampling $\boldsymbol{z}_t$, the state (i.e., the cluster) assignment at time t is changed, leading to updating $n_{i,j}^t$ accordingly. Consequently, the matrix Π is updated according to Equation 7.13.

Hyperparameter Sampling

In the HDP-HTM model, there are the concentration hyperparameters $\Theta = \{\alpha, \gamma, \lambda\}$. According to [52, 121], we may sample these parameters by the Gamma distribution with the constant Gamma parameters.

Finally, we summarize the EM framework as follows:

1. Initialize the transition matrix Π, as well as β, $\boldsymbol{m}$, and $\boldsymbol{z}_t$; compute $\boldsymbol{\omega}_t$ by taking the product of Π and β.

2. The E-step at time t:

   Sample the hyperparameters α, γ, and λ from the Γ distribution.

   Sample $\boldsymbol{z}_t$ based on Π, β, and α according to Equation 13.13.

   Sample $\boldsymbol{m}$ based on $\boldsymbol{z}_t$ and β according to Equation 13.15.

   Sample β based on $\boldsymbol{m}$ and γ according to Equation 13.16.

3. The M-step at time t:

   Update Π based on $\boldsymbol{z}_t$, β, and λ according to Equation 7.13.

4. Iterate 2 to 3 until convergence.

Part III

Applications

Chapter 14

Co-Clustering

In this Chapter, we apply our co-clustering algorithms to the application of document clustering to demonstrate that, as a general co-clustering algorithm, how NBVD improves the document clustering accuracy in comparison with NMF [36], and two other co-clustering algorithms, Information-theoretic Co-Clustering (ICC) [44] and Iterative Double Clustering (IDC) algorithm [49].

14.1   Data Sets and Implementation Details

We conduct the performance evaluation using the various subsets of 20-Newsgroup (*NG20*) data [82] and *CLASSIC3* data set [44]. The *NG20* data set consists of approximately 20,000 newsgroup articles collected evenly from 20 different usenet newsgroups. We have exactly duplicated this data set that is also used in [44,49] for document co-clustering in order to ensure the direct comparability in the evaluations. Many of the newsgroups share similar topics and about 4.5% of the documents are cross-posted making the boundaries between newsgroups rather fuzzy. To make our comparison consistent with the existing algorithms we have reconstructed various subsets of *NG*20 used in [44,49] to all the subsets, i.e., removing stop words, ignoring file headers, and selecting the top 2000 words based on the mutual information. As in [49], we include the subject line in the articles. Specific details of the subsets are given in Table 14.1.

The *CLASSIC3* [44] data set refers to the SMART collection from Cornell, which consists of MEDLINE, CISI, and CRANFIELD subcollections. The MEDLINE data consist of 1033 abstracts from medical journals; the CISI data consist of 1460 abstracts from information retrieval papers; and the CRANFIELD data consist of 1400 abstracts from the aerodynamic systems literature. We have preprocessed *CLASSIC3* similarly to what we have done for *NG20*, i.e., removing stop words and selecting the top 2000 words based on the mutual information.

Since each document vector of the word-by-document matrix is normalized to have unit L^2 norm in the implementation of NBVD, we normalize each column of RB to have unit L^2 norm. Assume that RB is normalized to RBV.

TABLE 14.1:   Data sets details. Each data set is randomly and evenly sampled from specific newsgroups

Data set Name	Newsgroups included	# Documents per group	Total documents
Binary	talk.politics.mideast, talk.politics.misc	250	500
Multi5	comp.graphics, rec.motorcycles, res.sports.baseball, sci.space, talk.politics.mideast	100	500
Multi10	alt.atheism, comp.sys.mac.hardware, misc.forsale, res.autos, res.sport.hockey, sci.crypt, sci.eldectronics, sci.med, sci.space, talk.politics.gun	50	500

The cluster labels for the documents are given by $V^{-1}C$ instead of C.

14.2   Evaluation Metricees

In the presence of the true labels, we can simply follow the accuracy measure used by the standard supervised learning algorithms. We measure the clustering performance using the accuracy given by the confusion matrix of the obtained clusters and the "real" classes. Each entry (i, j) in the confusion matrix represents the number of the documents in cluster i that are in true class j. Specifically, we use the micro-averaged-precision and the micro-averaged-recall as the evaluation measures, which are defined as follows:

$$P(\hat{c}) = \frac{\sum_c \alpha(c, \hat{c})}{\sum_c \alpha(c, \hat{c}) + \beta(c, \hat{c})} \tag{14.1}$$

$$R(\hat{c}) = \frac{\sum_c \alpha(c, \hat{c})}{\sum_c \alpha(c, \hat{c}) + \gamma(c, \hat{c})}, \tag{14.2}$$

where $\alpha(c, \hat{c})$ is the number of the documents correctly assigned to class c; $\beta(c, \hat{c})$ is the number of the documents incorrectly assigned to class c; and $\gamma(c, \hat{c})$ is the number of the documents belonging to class c but failed to be assigned to class c. If the data and the algorithms are both uni-labelled then $P(\hat{c}) = R(\hat{c})$.

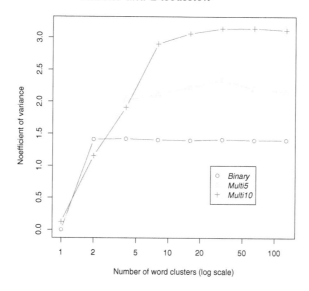

FIGURE 14.1: The coefficient of the variance for the columns of the mean block value matrix with the varing number of the word clusters using NBVD on different *NG20* data sets.

14.3   Results and Discussion

In the experiments, initial matrices are generated as follows. All the elements of R and C are generated from the uniform distribution between 0 and 1 and all the elements of B are simply assigned to the mean value of the data matrix. Since the NBVD algorithm is not guaranteed to find the global minimum, it is beneficial to run the algorithm several times with different initial values and choose one trial with the minimal objective value. In reality, usually a few number of trials is sufficient. In the experiments reported in this chapter, three trials of NBVD are performed in each test run and the final results are the averages for every twenty test runs. The experiments for NMF are conducted in the same way.

First, we demonstrate that the coefficient of the variance in the mean block value matrix may be used to indicate the optimal number of clusters in co-clustering. This is an "extra" advantage of the BVD framework because even for the traditional clusterings, the number of the clusters is not easy to determine. In the experiments for the *NG20* data set, we fix the document cluster number in this chapter as the number of the true document clusters and vary $k = 1, 2, \ldots, 128$ word clusters. Figure 14.1 documents how $CV(B_c)$ varies with the number of the word clusters for each data set. The minimal number of word clusters at which data set *Binary* reaches its maximal $CV(B_c)$ is 4.

TABLE 14.2: Both NBVD and NMF accurately recover the original clusters in the *CLASSIC3* data set

NBVD			NMF		
1008	1	2	1014	4	2
25	1459	19	18	1454	25
0	0	1379	1	2	1373

Hence, the optimal number of the word clusters for co-clustering *Binary* is 4. Similarly, the optimal number of the word clusters for data sets *Multi5* and *Multi10* are 32 and 64, respectively. In general selecting the optimal number of clusters is a nontrivial model selection problem and is beyond the scope of this paper. Figure 14.2 shows how the precision values vary with the num-

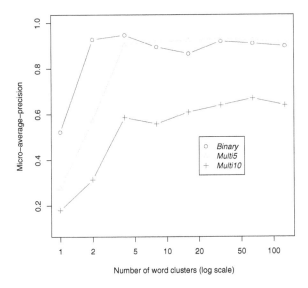

Number of word clusters (log scale)

FIGURE 14.2: Micro-averaged-precision with the varing number of the word clusters using NBVD on different *NG20* data sets.

ber of word clusters for each data set. The curves have the similar patterns to those of the curves in Figure 14.1. *Binary* reaches its maximal precision at 4-word clusters; *Multi5* reaches its maximal precision at 32-word clusters; and *Multi10* reaches its maximal precision at 64-word clusters. Clearly, the optimal numbers of the word clusters found in both scenarios in Figures 14.2 and 14.1, respectively, are completely consistent.

Table 14.2 records the two confusion matrices obtained on *CLASSIC3* data

TABLE 14.3: A normalized block value matrix on the *CLASSIS3* data set

0.701	0.000	0.000
0.000	0.608	0.000
0.000	0.000	1.000

TABLE 14.4: NBVD extracts the block structure more accurately than NMF on *Multi5* data set

NBVD					NMF				
92	1	4	3	1	94	4	4	13	2
2	96	3	3	0	1	88	5	5	4
1	0	93	1	0	2	3	90	5	2
4	1	0	93	1	3	4	1	77	0
1	2	0	0	98	0	1	0	3	93

set using NMF and NBVD, respectively, with 3-word clusters that is the number of the true word clusters. Observe that NBVD has extracted the original clusters with micro-averaged-precision of 0.9879 and NMF has led to a micro-averaged-precision of 0.9866. It is not surprising that NBVD and NMF have almost the same performance on *CLASSIC3* data set. This is due to the fact that when there exists a perfect one-to-one correspondence between row clusters and column clusters, the block value matrix B is close to the identity matrix and NMF is equivalent to NBVD. Table 14.3 shows a block value matrix for data set *CLASSIC3*. The perfect diagonal structure of Table 14.3 indicates the one-to-one correspondence structure between the document clusters and the word clusters for *CLASSIC3*.

Table 14.4 shows the two confusion matrices obtained on the *Multi5* data set by NBVD and NMF, respectively. NBVD and NMF yield micro-averaged-precision of 0.944 and 0.884, respectively. This experiment shows that NBVD has a better performance than NMF on the data set *Multi5*. Compared with *CLASSIC3*, *Multi5* has more complicated hidden block structure and there is no simple one-to-one relationship between the document clusters and the word clusters. This demonstrates that by exploiting the duality of the row clustering and the column clustering, NBVD is more powerful to discover the complicated hidden block structures of the data than NMF.

Table 14.5 shows the micro-averaged-precision measures on all the data sets from *NG20* data. All NBVD precision values are obtained by running NBVD on the corresponding optimal numbers of the word clusters from Figure 11.2. The peaked ITC and IDC precision values are quoted from [44] and [49], respectively. On all data sets NBVD performs better than its one-sided counterpart NMF. This result justifies the need to exploit the duality between the word clustering and the document clustering. Compared with other two

TABLE 14.5: NBVD shows clear improvements on the micro-averaged-precision values on different newsgroup data sets over other algorithms

	NBVD	NMF	ICC	IDC
Binary	0.95	0.91	0.96	0.85
Multi5	0.93	0.88	0.89	0.88
Multi10	0.67	0.60	0.54	0.55

state-of-the-art co-clustering algorithms, NBVD shows clear improvements on precision for almost all the data sets. In particular, more substantial improvements are observed on the complicated data sets with more clusters, which is the typical scenario in practice.

Chapter 15

Heterogeneous Relational Data Clustering

In this chapter, we provide application examples for heterogeneous relational data clustering. In particular, we apply RSN-BD to two basic types of k-partite heterogeneous relation graphs: the bipartite graph and the sandwich structure tripartite graph (such as Figure 3.2a), which arise frequently in various applications. Note that the application of RSN-BD is not only limited to these two types of graphs but also applicable to various k-partite heterogeneous relation graphs. Four types of RSN-BD are evaluated in the experiments: RSN with Euclidean Distance (RSN-ED) assumes the normal distribution of the data; RSN with Logistic Loss (RSN-LL) assumes the Bernoulli distribution of the data; RSN with Generalized I-divergence (RSN-GI) assumes the Poisson distribution of the data; RSN with Itakura-Saito distance (RSN-IS) assumes the exponential distribution of the data. Two graph partitioning approaches, BSGP [43] and Consistent Bipartite Graph Co-partitioning (CBGC) [56] (we thank the authors for providing the executable code of CBGC) are used as the comparison on bipartite graph and sandwich tripartite graph, respectively. Four traditional feature-based algorithms, which cluster a type of nodes in a k-partite heterogeneous relation graph by transforming all the links into features, are also used as comparisons. They are K-Means with Euclidean Distance (KM-ED), K-Means with Logistic Loss (KM-LL), K-Means with Generalized I-divergence (KM-GI) and K-Means with Itakura-Saito (KM-IS).

15.1 Data Sets and Parameter Setting

The data sets used in the experiments include synthetic data sets with various distributions and real data sets based on the 20-Newsgroup data [?].

The synthetic bipartite graphs are generated such that both V_1 and V_2 have two clusters (to be fair for BSGP, we use an equal number of clusters); each cluster has 100 nodes, hence, both V_1 and V_2 have 200 nodes. The distributions and parameters (the true means of the distributions) used to generate the links in the graphs are documented in Table 15.1. In the table, distribution parameters for a graph is represented as a matrix S such that S_{pq} denotes

TABLE 15.1:   Parameters and distributions for synthetic bipartite graphs

Data set	S	Distribution
BP-b1	0.1 0.9 0.9 0.1	Bernoulli
BP-b2	0.4 0.7 0.5 0.6	Bernoulli
BP-p	0.5 0.6 0.6 0.8	Poisson
BP-e	0.4 0.5 0.5 0.7	Exponential

TABLE 15.2:   Subsets of newsgroup data for constructing bipartite graphs

Dataset Name	Newsgroups Included	# Documents per Group	Total # Documents
BP-NG1	rec.sport.baseball, rec.sport.hockey	200	400
BP-NG2	comp.os.ms-windows.misc, comp.windows.x, rec.motorcycles, sci.crypt, sci.space	200	1000
BP-NG3	comp.os.ms-windows.misc, comp.windows.x, misc.forsale, rec.motorcycles, rec.motorcycles,sci.crypt, sci.space, talk.politics.mideast, talk.religion.misc	200	1600

the mean parameter of the distribution to generate the links between the pth cluster of V_1 and the qth cluster of V_2.

The real bipartite graphs are constructed based on various subsets of the 20-Newsgroup data [?] which contains about 20,000 articles from 20 newsgroups. We preprocess the data by removing stop words and selecting the top 2000 words by the mutual information. The document-word matrix is based on *tf.idf* weighting scheme and each document vector is normalized to a unit L_2 norm vector. Specific details of data sets used to construct bipartite graphs are listed in Table 15.2. For example, to construct a *BP-NG3* graph, we randomly and evenly sample 200 documents from the corresponding newsgroups; then we formulate a bipartite graph, consisting of 1600 document nodes and 2000 word nodes.

The synthetic tripartite graphs are generated similarly to the bipartite graphs. The distributions and parameters are documented in Table 15.3. Let V_1 denote the central type nodes. In Table 15.3, $S^{(12)}$ denotes the true means of distributions for generating the links between V_1 and V_2, and similarly for $S^{(13)}$. The numbers of clusters for each type of nodes are given by dimensions

TABLE 15.3: Parameters and distributions for synthetic tripartite graphs

Data set	$S^{(12)}$	$S^{(13)}$	Distribution
TP-b1	0.4 0.7 0.5 0.6	0.7 0.5 0.6 0.6	Bernoulli
TP-b2	0.5 0.6 0.5 0.6 0.7 0.7	0.6 0.6 0.6 0.7 0.7 0.5 0.7 0.7 0.5	Bernoulli
BP-p	0.3 0.6 0.2 0.7	0.4 0.4 0.5 0.3	Poisson
TP-large	$\mathbb{Z}^{20\times20}$	$\mathbb{Z}^{20\times18}$	Poisson
TP-e	0.3 0.6 0.3 0.7	0.4 0.7 0.5 0.6	Exponential

TABLE 15.4: Taxonomy structures of two data sets for constructing tripartite graphs

Data set	Taxonomy structure
TP-TM1	{rec.sport.baseball, rec.sport.hockey}, {talk.politics.guns, talk.politics.mideast, talk.politics.misc}
TP-TM2	{comp.graphics, comp.os.ms-windows.misc}, {rec.autos, rec.motorcycles}, {sci.crypt, sci.electronics}

of $S^{(12)}$ and $S^{(13)}$ and each cluster has 100 nodes. In Table 15.3, *TP-large* is a large graph with 20 clusters of V_1, 20 clusters of V_2, and 18 clusters of V_3 (due to the space limit, the details of parameters are omitted). Each *BP-large* graph contains 5800 nodes and on an average about 3.25 million links.

The real tripartite graphs are built based on the 20-newsgroups data for hierarchical taxonomy mining. In the field of text categorization, hierarchical taxonomy classification is widely used to obtain a better trade-off between effectiveness and efficiency than flat taxonomy classification. To take advantage of hierarchical classification, one must mine a hierarchical taxonomy from the data set. We see that words, documents, and categories formulate a sandwich structure tripartite graph, in which documents are central type nodes. The links between documents and categories are constructed such that if a document belongs to k categories, the weights of the links between this document and these k category nodes are $1/k$ (please refer [56] for details).

The true taxonomy structures for two data sets, *TP-TM1* and *TP-TM2*, are documented in Table 15.4. For example, *TP-TM1* data set is sampled from five categories (200 documents for each category) in which two categories belong to the high-level category *res.sports* and other three categories belong to the high-level category *talk.politics*.

For all the algorithms on all the graphs, we fix the number of iterations to 20 (this also holds true for BSGP and CBGC, since they use classic k-means to do postprocessing) and use the same initialization, a random initialization for synthetic data and the classic k-means initialization for real data. The final performance score is the average of 20 runs. At each test run, a graph is constructed by sampling from the corresponding distributions or newsgroups of the 20-newsgroup data. Hence, the variation of a final performance score includes the variance of sampling.

For the number of clusters, we use the true number of clusters for the synthetic graphs. For real data graphs, we use the true number of clusters for documents and categories; however, we do not know the true number of word clusters. How to determine the optimal number of word clusters is beyond the scope of this research. We simply adopt 40 for all the RSN algorithms. For BSGP and CBGC, the number of word clusters must equal the number of document clusters. By the authors' suggestion, the parameter setting for CBGC is $\beta = 0.5$, $\theta_1 = 1$, and $\theta_2 = 1$ [56].

The performance comparison is based on the quality of the clusters of one type of nodes in each graph. In synthetic bipartite graphs, it is based on V_1 whose clusters correspond to the rows of S in Table 15.1; in synthetic tripartite graphs, it is based on the central type nodes V_1; in bipartite graphs of documents and words, it is based on the documents; in tripartite graphs for taxonomy mining, it is based on the categories whose clusters provide the taxonomy structures. For the performance measure, we elect to use the Normalized Mutual Information (NMI) [119], which is a standard way to measure the cluster quality.

15.2   Results and Discussion

Table 15.5 shows the NMI scores of the nine algorithms on the bipartite graphs. For the *BP-b1* graph, all the algorithms provide the perfect NMI score, since the graphs are generated with very clear structures, which can be seen from the parameter matrix in Table 15.1. For other synthetic bipartite graphs, the cluster structures are subtle, especially for the nodes V_1, whose cluster structures are our objective. For these graphs, the RSN algorithms perform much better than k-means algorithms, especially for the *BP-b2* and *BP-p* graph, in which the distributions for clusters of V_1 are very close to each other and the links are relatively sparse. This comparison implies that benefiting from the interactions among the cluster structures of different types of nodes, the RSN algorithms are able to identify very subtle cluster structures even when the traditional clustering approaches totally fail. Compared with the RSN algorithms, BSGP performs poorly for all the synthetic bipartite graphs except *BP-b1*. The possible explanation is that it assumes one-to-one associations between clusters of different types of nodes, which does not hold

TABLE 15.5: NMI scores of the algorithms on bipartite graphs

Algorithm	*BP-b1*	*BP-b2*	*BP-p*	*BP-e*	*BP-NG1*	*BP-NG2*	*BP-NG3*
RSN-ED	1	0.618	0.549	0.821	0.402	0.599	0.573
KM-ED	1	0.069	0.042	0.632	0.375	0.616	0.601
RSN-LL	1	**0.620**	0.519	0.819	**0.638**	**0.747**	**0.698**
KM-LL	1	0.060	0.224	0.567	0.443	0.655	0.641
RSN-GI	1	0.604	**0.562**	0.849	0.619	0.746	0.697
KM-GI	1	0.053	0.025	0.656	0.444	0.655	0.641
RSN-IS	1	0.549	0.553	**0.857**	0.411	0.414	0.335
KM-IS	1	0.050	0.025	0.635	0.383	0.618	0.596
BSGP	1	0.379	0.005	0.004	0.430	0.638	0.501

true for the synthetic bipartite graphs except *BP-b1*. We also observe that the RSN algorithm with the distance function matching the distribution to generate the graph provides the best NMI score for that graph.

For the real bipartite graphs consisting of document and word nodes, RSN-LL always provides the best NMI score. For the difficult *BP-NG1* graph based on two "close" newsgroups, RSN-LL shows about 44% improvement in comparison with KM-LL, which is, along with KM-GI, the best among the non-RSN algorithms. Note that since the document vector is L_2-normalized, the KM-ED is actually based on von Mises-Fisher distribution [91], which proved efficient for document clustering [9]. We also observe that for these graphs, in general the algorithms based on logistic loss provide better performance. The possible reason is that logistic loss corresponds to Bernoulli distribution which provides a good approximation to the distribution of the data consisting of a large number of zeros, such as the sparse links between documents and words. In the meantime, it is also reasonable to assume the Poisson distribution for the frequency that a word appears in a document. That is why RSN-GI also shows the performance very close to RSN-LL. The above comparison verifies the assumption that under an appropriate distribution assumption, through the hidden nodes the RSN algorithms perform implicit adaptive feature reduction to overcome the typical high dimensionality and sparseness.

Table 15.6 shows the NMI scores of the nine algorithms on the tripartite graphs. As similarly in the synthetic bipartite graphs, the RSN algorithms perform much better than the k-means algorithms. Except for RSN-ED on the *TP-p* graph, the RSN algorithms perform significantly better than CBGC. The NMI scores of CBGC for some graphs are not available because the CBGC code provided by the authors only works for the case of two clusters and small-size graphs. For the large dense *TP-large* graph, the RSN algorithms perform consistently better than the KM algorithms, and this demonstrates the good scalability of the RSN algorithms; RSN-ED performs the best on *TP-large*, and this demonstrates the advantage of the normal distribution for the very large sample size of dense links.

For the real tripartite graphs for taxonomy mining, the k-means algorithms

TABLE 15.6: NMI scores of the algorithms on tripartite graphs

Algorithm	TP-b1	TP-b2	TP-p	TP-large	TP-e	TP-TM1	TP-TM2
RSN-ED	0.835	0.847	0.573	**0.715**	0.612	**0.887**	0.623
KM-ED	0.196	0.258	0.012	0.165	0.017	0.257	0.439
RSN-LL	**0.848**	**0.860**	0.622	0.335	0.606	0.858	0.645
KM-LL	0.219	0.255	0.025	0.174	0.016	0.218	0.456
RSN-GI	0.829	0.854	**0.656**	0.658	0.662	0.858	0.637
KM-GI	0.194	0.289	0.014	0.174	0.019	0.245	0.482
RSN-IS	0.801	0.811	0.616	0.512	**0.677**	**0.887**	**0.681**
KM-IS	0.152	0.3100	0.019	0.250	0.012	0.223	0.469
CBGC	0.744	—	0.575	—	0.575	—	—

perform poorly since they cluster categories only based on links between categories and documents. From Table 15.6, we observe that both RSN-ED and RSN-IS provide the best NMI score for *TP-TM1*. To have an intuition about this score, we check the details of the 20 test runs, which show that in 16 out of the 20 runs the algorithms provide the perfect taxonomy structures and in the other 4 runs one category is clustered incorrectly.

Chapter 16

Homogeneous Relational Data Clustering

In this chapter, we present application examples for the homogeneous relational data clustering algorithms, CLGA algorithms.

16.1   Data Sets and Parameter Setting

The data sets used in the experiments include synthetic graphs with different types of cluster structures and real graphs for text mining and social network analysis.

The synthetic graphs are 0-1 graphs generated based on Bernoulli distribution. The distribution parameters to generate the graphs are listed in the second column of Table 16.1 as matrices. In a parameter matrix P, P_{ij} denotes the probability that the nodes in the ith cluster are connected to those in the jth cluster. For example, in graph W1, the nodes in cluster 1 are connected to those in cluster 2 with probability 0.1 and the nodes within clusters are connected to each other with probability 0. The graph G2 has 10 clusters mixing with strongly and weakly intra-connected clusters. Due to the space limit, its distribution parameters are omitted here. Totally G2 has 5000 nodes and about 2.1 million edges.

The graphs based on the text data have been widely used to test graph learning algorithms [43,47,69]. We use various data sets from the 20-newsgroups [?], WebACE, and TREC to construct the real graphs. The data are preprocessed by removing the stop words and each document is represented by a term-frequency vector using TF-IDF weights. Then we construct a graph for each data set such that each node denotes a document and the edge weight denotes the cosine similarity between documents. A summary of all the data sets to construct graphs used in this chapter is shown in Table 16.2 in which n denotes the number of nodes in a graph, k denotes the number of true clusters, and *balance* denotes the size ratio of the smallest cluster to the largest cluster. Besides the graphs constructed directly from these data sets, we also construct three graphs with virtual nodes, v-tr23, v-tr45, and v-NG1-20, for three relatively difficult data sets. To incorporate supervised information, $5\%n$ virtual

TABLE 16.1: Summary of the graphs with general clusters

Graph	Parameter	n	k
S1	$\begin{bmatrix} 0.6\ 0.3\ 0.3 \\ 0.3\ 0.5\ 0.3 \\ 0.3\ 0.3\ 0.5 \end{bmatrix}$	1500	3
W1	$\begin{bmatrix} 0\ \ 0.1\ 0.1 \\ 0.1\ \ 0\ \ 0.3 \\ 0.1\ 0.3\ \ 0 \end{bmatrix}$	1500	3
G1	$\begin{bmatrix} 0.3\ 0.2\ 0.3 \\ 0.2\ \ 0\ \ 0.2 \\ 0.3\ 0.2\ \ 0 \end{bmatrix}$	1500	3
G2	$[0,1]^{10 \times 10}$	5000	10

TABLE 16.2: Summary of graphs based on text datasets

Name	n	k	Balance	Source
tr11	414	9	0.046	TREC
tr23	204	6	0.066	TREC
tr45	690	10	0.0856	TREC
NG1-3	1600	3	0.5	20-newsgroups
NG1-20	14000	20	1.0	20-newsgroups
k1b	2340	6	0.043	WebACE

nodes are added into the original graphs. To simulate the concept documents provided by domain expert in real applications, the "mean" documents of the clusters, which contain popular words of the corresponding clusters, are used as virtual documents.

For the performance measure, we elect to use the Normalized Mutual Information (NMI) [119] between the resulting cluster labels and the true cluster labels, which is a standard way to measure the cluster quality. For the number of clusters k, we simply use the number of the true clusters, since how to choose the optimal number of clusters is a nontrivial model selection problem and beyond the scope of this study.

Three versions of the algorithms under different CLGA models as listed in Table 4.1 are tested in the experiments. We use "H-", "S-" and "B-" to represent hard, soft, and balanced versions, respectively. Two representative graph learning algorithms are selected in the comparisons. Spectral approaches have been applied to a wide range of graph learning tasks from regular graph to bipartite [43] and k-partite graph learning [87]. In this study, we select to use Spectral Graph Clustering (SGC) [96] that is a generalization of a number of graph learning algorithms. SGC learns relation-pattern-based cluster structures by embedding cluster structures into eigen-space and discovering them by the eigenvectors. Another comparison algorithm is the classic multilevel graph partitioning algorithm, METIS [75].

TABLE 16.3: NMI scores on graphs of general clusters

Algorithm	S1	W1	G1	G2
SGC	0.9902	0.4551	0.5086	0.6125
METIS	0.9811	0.0108	0.1495	0.6391
H-GGP	**1.0000**	0.0582	0.0151	0.6965
H-GBCL	0.0031	**1.000**	0.7932	0.8837
H-GCL	0.9103	0.9306	0.6531	**0.9133**
S-GGP	**1.0000**	0.0006	0.0009	0.5802
S-GBCL	0.0025	0.9270	0.8100	0.6494
S-GCL	0.9211	0.9295	0.8144	0.7314
B-GGP	**1.0000**	0.0002	0.0008	0.5832
B-GBCL	0.0011	0.9091	0.9976	0.7128
B-GCL	0.9666	0.9085	**1.0000**	0.7695

16.2   Results and Discussion

Table 16.3 shows the NMI scores of the 11 algorithms on the graphs listed in Table 16.1. Each NMI score is the average of 10 test runs. The graph S1 has three strongly intra-connected clusters. We observe that all the GGP algorithms provide perfect scores on S1. However, the GBCL algorithms totally fail, since their goal is to learn weakly intra-connected clusters which do not exist in S1. The GCL algorithms do not perform as well as the GGP algorithms. The possible reason is that they do not focus on strongly intra-connected clusters and hence have more local optimal solutions to converge. The graph W1 consists of three weakly intra-connected clusters. This time the GBCL algorithms provide perfect or nearly perfect performance scores. METIS totally fails, since it only looks for strongly intra-connected clusters. SGC identifies part of cluster structures, but its performance is not satisfactory. Both G1 and G2 are graphs of mixed-type cluster structures. Since there is only one strongly intra-connected cluster in G1, it is still difficult for the GGP algorithms and METIS to identify it. The graph G2 is a large graph of 10 clusters consisting of 4 strongly intra-connected clusters and 6 weakly intra-connected clusters. The GGP and METIS algorithms do not totally fail on G2, since they could learn strongly intra-connected part from G2. The GBCL algorithms could learn weakly intra-connected part from G2. However, overall the GCL algorithms perform better especially on G1 and G2, since they are capable of learning general clusters of various types. In summary, the CLGA model provides a family of algorithms for effectively learning general relation-pattern-based clusters as well as specific types of clusters.

Table 16.4 shows the NMI scores of the nine algorithms. Since the graphs mainly consist of strongly intra-connected clusters, GBCL algorithms are not appropriate. Hence IGP (GGP works similarly to IGP and their results are

TABLE 16.4: NMI scores on graphs of text data

Data	VMF	SGC	METIS	H-IGP	H-GCL	S-IGP	S-GCL	B-IGP	B-GCL
tr11	0.5990	0.5848	0.5708	0.5902	0.6048	**0.6070**	0.5941	0.6016	0.5959
tr23	0.2420	0.3003	0.2454	**0.3223**	0.2158	0.3065	0.3011	0.2905	0.2894
tr45	0.4821	0.5224	0.4290	**0.5570**	0.4991	0.4753	0.4793	0.5068	0.5107
NG1-3	0.5611	0.5515	0.4998	0.4574	0.5101	**0.6132**	0.6104	0.5840	0.5799
NG1-20	0.5102	0.4770	**0.5343**	0.4977	0.4015	0.5131	0.5102	0.5124	0.5211
k1b	0.4998	0.4758	0.5058	0.5066	0.5007	**0.5146**	0.5087	0.4946	0.4903
v-tr23	0.4001	0.3805	0.3530	0.3873	0.4147	0.4606	**0.4648**	0.4265	0.4492
v-tr45	0.6095	0.6225	0.6050	0.6184	0.6205	0.6163	**0.6304**	0.5953	0.5967
v-NG1-20	0.5468	0.5597	0.5684	0.5873	0.5954	0.6031	0.6208	0.6177	**0.6336**

omitted) and GCL results are reported in comparison with SGC, METIS, and a state-of-the-art document clustering algorithm, VMF, which is based on von Mises-Fisher distribution and was reported as one of the best document clustering algorithm [10]. We observe that although there is no single winner on all the graphs, for most graphs CLGA algorithms perform better than or close to SGC, METIS, and VMF. The CLGA algorithms provide the best performance on eight out of the nine graphs. By adding virtual nodes into the graphs, the cluster structures are reinforced by relation patterns induced by the virtual nodes. Although all the algorithms benefit from the virtual nodes, CLGA algorithms always achieve the best performance on the graphs with virtual nodes. For example, S-GCL provides the best performance on v-tr23 and increases the performance about 50% by making use of the virtual nodes. Hence, when learning the clusters of documents (with or without supervised information), the CLGA model provides a family of new algorithms which are competitive compared with the existing state-of-the-art graph learning algorithms and the document clustering algorithm.

Chapter 17

General Relational Data Clustering

In this chapter, we show several applications for the general relational data clustering algorithm, MMRC algorithm. Since a number of state-of-the-art clustering algorithms [11,13,14,31,44,85] can be viewed as special cases of EF-MMRC model and algorithms, the experimental results in these efforts also illustrate the effectiveness of the MMRC model and algorithms. In this chapter, we apply MMRC algorithms to tasks of graph clustering, bi-clustering, tri-clusering, and clustering on a general relational data set of all three types of information. In the experiments, we use mixed version MMRC, i.e., hard MMRC initialization followed by soft MMRC. Although MMRC can adopt various distribution assumptions, due to space limit, we use MMRC under normal or Poisson distribution assumption in the experiments. However, this does not imply that they are optimal distribution assumptions for the data. How to decide the optimal distribution assumption is beyond the scope of this study.

For performance measure, we elect to use the Normalized Mutual Information (NMI) [119] between the resulting cluster labels and the true cluster labels, which is a standard way to measure the cluster quality. The final performance score is the average of 10 runs.

17.1 Graph Clustering

In this section, we present experiments on the MMRC algorithm under normal distribution in comparison with two representative graph partitioning algorithms, the spectral graph partitioning (SGP) from [96] that is generalized to work with both normalized cut and ratio association, and the classic multilevel algorithm, METIS [75].

The graphs based on the text data have been widely used to test graph partitioning algorithms [43, 47, 69]. In this study, we use various data sets from the 20-Newsgroups [?], WebACE, and TREC, which cover data sets of different sizes, different balances, and different levels of difficulties. The data are preprocessed by removing the stop words and each document is represented by a term-frequency vector using the TF-IDF weights. Then we

TABLE 17.1: Summary of relational data for Graph Clustering

Name	n	k	Balance	Source
tr11	414	9	0.046	TREC
tr23	204	6	0.066	TREC
NG1-20	14000	20	1.0	20-newsgroups
k1b	2340	6	0.043	WebACE

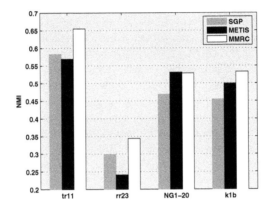

FIGURE 17.1: NMI comparison of SGP, METIS, and MMRC algorithms.

construct relational data for each text data set such that objects (documents) are related to each other with cosine similarities between the term-frequency vectors. A summary of all the data sets to construct relational data used in this chapter is shown in Table 17.1, in which n denotes the number of objects in the relational data, k denotes the number of true clusters, and *balance* denotes the size ratio of the smallest clusters to the largest clusters.

For the number of clusters k, we simply use the true number of the clusters. Note that how to choose the optimal number of clusters is a nontrivial model selection problem and beyond the scope of this study.

Figure 17.1 shows the NMI comparison of the three algorithms. We observe that although there is no single winner on all the graphs, overall the MMRC algorithm performs better than SGP and METIS . Especially on the difficult data set tr23, MMRC increases performance about 30%. Hence, MMRC under normal distribution provides a new graph partitioning algorithm which is viable and competitive compared with the two existing state-of-the-art graph partitioning algorithms. Note that although the normal distribution is most popular, MMRC under other distribution assumptions may be more desirable in specific graph clustering applications, depending on the statistical properties of the graphs.

TABLE 17.2: Subsets of newsgroup data for bi-type relational data

Data set Name	Newsgroups Included	# Documents per group	Total # documents
BT-NG1	rec.sport.baseball, rec.sport.hockey	200	400
BT-NG2	comp.os.ms-windows.misc, comp.windows.x, rec.motorcycles, sci.crypt, sci.space	200	1000
BT-NG3	comp.os.ms-windows.misc, comp.windows.x, misc.forsale, rec.motorcycles, rec.motorcycles, sci.crypt, sci.space, talk.politics.mideast, talk.religion.misc	200	1600

TABLE 17.3: Taxonomy structures of two data sets for constructing tripartite relational data

Data set	Taxonomy structure
TT-TM1	{rec.sport.baseball, rec.sport.hockey}, {talk.politics.guns, talk.politics.mideast, talk.politics.misc}
TT-TM2	{comp.graphics, comp.os.ms-windows.misc}, {rec.autos, rec.motorcycles}, {sci.crypt, sci.electronics}

17.2 Bi-clustering and Tri-Clustering

In this section, we apply the MMRC algorithm under Poisson distribution to clustering bi-type relational data, word-document data, and tri-type relational data, word-document-category data. Two algorithms, Bipartite Spectral Graph Partitioning (BSGP) [43] and Relation Summary Network under Generalized I-divergence (RSN-GI) [86], are used in comparison for bi-clustering. For tri-clustering, Consistent Bipartite Graph Co-partitioning (CBGC) [56] and RSN-GI are used in comparison.

The bi-type relational data, word-document data, are constructed based on various subsets of the 20-newsgroup data. We preprocess the data by selecting the top 2000 words by the mutual information. The document-word matrix is based on *tf.idf* weighting scheme and each document vector is normalized to a unit L_2 norm vector. Specific details of the data sets are listed in Table 17.2. For example, for the data set *BT-NG3* we randomly and evenly sample 200 documents from the corresponding newsgroups; then we formulate a bi-type relational data set of 1600 document and 2000 word.

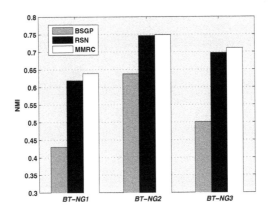

FIGURE 17.2: NMI comparison among BSGP, RSN, and MMRC algorithms for bi-type data.

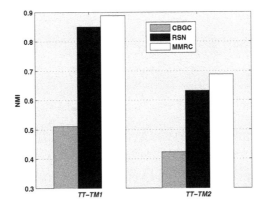

FIGURE 17.3: NMI comparison of CBGC, RSN, and MMRC algorithms for tri-type data.

TABLE 17.4: Two clusters from actor-movie data

Cluster 23 of actors
Viggo Mortensen, Sean Bean, Miranda Otto, Ian Holm, Christopher Lee, Cate Blanchett, Ian McKellen ,Liv Tyler , David Wenham , Brad Dourif , John Rhys-Davies , Elijah Wood , Bernard Hill , Sean Astin , Andy Serkis , Dominic Monaghan , Karl Urban , Orlando Bloom , Billy Boyd ,John Noble, Sala Baker
Cluster 118 of movies
The Lord of the Rings: The Fellowship of the Ring (2001) The Lord of the Rings: The Two Towers (2002) The Lord of the Rings: The Return of the King (2003)

The tri-type relational data are built based on the 20-newsgroups data for hierarchical taxonomy mining. In the field of text categorization, hierarchical taxonomy classification is widely used to obtain a better trade-off between effectiveness and efficiency than flat taxonomy classification. To take advantage of hierarchical classification, one must mine a hierarchical taxonomy from the data set. We see that words, documents, and categories formulate a sandwich structure tri-type relational data set, in which documents are the central type nodes. The links between documents and categories are constructed such that if a document belongs to k categories, the weights of the links between this document and these k category nodes are $1/k$ (please refer [56] for details). The true taxonomy structures for two data sets, *TP-TM1* and *TP-TM2*, are documented in Table 17.3.

Figures 17.2 and 17.3 show the NMI comparisons of the three algorithms on bi-type and tri-type relational data, respectively. We observe that the MMRC algorithm performs significantly better than BSGP and CBGC. MMRC performs slightly better than RSN on some data sets. Since RSN is a special case of hard MMRC, this shows that mixed MMRC improves hard MMRC's performance on the data sets. Therefore, compared with the existing state-of-the-art algorithms, the MMRC algorithm performs more effectively on these bi-clustering and tri-clustering tasks, and, on the other hand, it is flexible for different types of multi-clustering tasks which may be more complicated than tri-type clustering.

17.3   A Case Study on Actor-Movie Data

We also run the MMRC algorithm on the actor-movie relational data based on the IMDB movie data set for a case study. In the data, actors are related to each other by collaboration (homogeneous relations); actors/actresses are related to movies by taking roles in movies (heterogeneous relations); movies have attributes such as release time and rating (note that there are no links between movies). Hence the data have all the three types of information. We formulate a data set of 20,000 actors/actresses and 4000 movies. We run experiments with $k = 200$. Although there is no ground truth for the data's cluster structure, we observe that most resulting clusters that are actors or movies of the similar style such as action, or tight groups from specific movie serials. For example, Table 17.4 shows cluster 23 of actors and cluster 118 of movies; the parameter $\Upsilon_{23,118}$ shows that these two clusters are strongly related to each other. In fact, the actors/actresses cluster contains the actors/actresses in the movie series "The Lord of the Rings." Note that if we only have one type of actors/actresses objects, we only get the actors/actresses clusters, but with two types of nodes, although there are no links between the movies, we also get the related movie clusters to explain how the actors are related.

17.4   Spectral Relational Clustering Applications

In this section, we evaluate the effectiveness of the SRC algorithm on two types of relational data: bi-type relational data and tri-type star-structured data.

The data sets used in the experiments are mainly based on the 20-Newsgroup data [?] which contains about 20,000 articles from 20 newsgroup. We pre-process the data by removing stop words and file headers and selecting top 2000 words by the mutual information. The word-document matrix R is based on *tf.idf* and each document vector is normalized to the unit norm vector. In the experiments, the classis k-means is used for initialization and the final performance score for each algorithm is the average of the 20 test runs unless stated otherwise.

17.4.1   Clustering on Bi-Type Relational Data

In this section, we conduct experiments on a bi-type relational data, word-document data, to demonstrate the effectiveness of SRC as a novel co-clustering algorithm. A representative spectral clustering algorithm, Normalized-Cut (NC) spectral clustering [96, 113], and BSGP [43], are used as comparisons.

The graph affinity matrix for NC is $R^T R$, i.e., the cosine similarity matrix. In NC and SRC, the leading k eigenvectors are used to extract the cluster structure, where k is the number of document clusters. For BSGP, the second to the $(\lceil \log_2 k \rceil + 1)$th leading singular vectors are used [43]. K-means

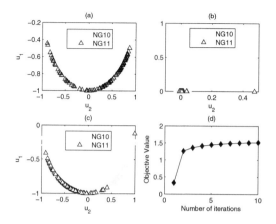

FIGURE 17.4: (a), (b), and (c) are document embeddings of multi2 data set produced by NC, BSGP, and SRC, respectively (u_1 and u_2 denote first and second eigenvectors, respectively). (d) is an iteration curve for SRC.

is adopted to postprocess the eigenvectors. Before postprocessing, the eigenvectors from NC and SRC are normalized to the unit norm vector and the eigenvectors from BSGP are normalized as described by [43]. Since all the algorithms have random components resulting from k-means or itself, at each test we conduct three trials with random initializations for each algorithm and the optimal one provides the performance score for that test run. To evaluate the quality of document clusters, we elect to use the Normalized Mutual Information (NMI) [119], which is a standard way to measure the cluster quality.

At each test run, five data sets, multi2 (NG 10, 11), multi3(NG 1, 10, 20), multi5 (NG 3, 6, 9, 12, 15), multi8 (NG 3, 6, 7, 9, 12, 15, 18, 20), and multi10 (NG 2, 4, 6, 8, 10, 12, 14, 16, 18, 20), are generated by randomly sampling 100 documents from each newsgroup. Here NG i means the ith newsgroup in the original order. For the numbers of document clusters, we use the numbers of the true document classes. For the numbers of word clusters, there are no options for BSGP, since they are restricted to equal to the numbers of document clusters. For SRC, it is flexible to use any number of word clusters. Since how to choose the optimal number of word clusters is beyond the scope of this paper, we simply choose one more word clusters than the corresponding document clusters, i.e., 3,4, 6, 9, and 11. This may not be the best choice, but it is good enough to demonstrate the flexibility and effectiveness of SRC.

In Figure 17.4a, b, and c shows three document embeddings of a multi2 sample, which is sampled from two close newsgroups: *rec.sports.baseball* and *rec.sports.hockey*. In this example, when NC and BSGP fail to separate the document classes, SRC still provides a satisfied separation. The possible explanation is that the adaptive interactions among the hidden structures of word clusters and document clusters remove the noise to lead to better embeddings.

TABLE 17.5:   NMI comparisons of SRC, NC, and BSGP algorithms

Data set	SRC	NC	BSGP
multi2	0.4979	0.1036	0.1500
multi3	0.5763	0.4314	0.4897
multi5	0.7242	0.6706	0.6118
multi8	0.6958	0.6192	0.5096
multi10	0.7158	0.6292	0.5071

Figure 17.4d shows a typical run of the SRC algorithm. The objective value is the trace value in Theorem 12.3.

Table 17.5 shows NMI scores on all the data sets. We observe that SRC performs better than NC and BSGP on all data sets. This verifies the hypothesis that benefiting from the interactions of the hidden structures of different types of objects, the SRC's adaptive dimensionality reduction has advantages over the dimensionality reduction of the existing spectral clustering algorithms.

17.4.2   Clustering on Tri-Type Relational Data

In this section, we conduct experiments on tri-type star-structured relational data to evaluate the effectiveness of SRC in comparison with other two algorithms for MTRD clustering. One is based on m-partite graph partitioning, Consistent Bipartite Graph Co-partitioning (CBGC) [56] (we thank the authors for providing the executable program of CBGC). The other is Mutual Reinforcement K-means (MRK), which is implemented based on the idea of mutual reinforcement clustering [137].

The first data set is synthetic data, in which two relation matrices, $R^{(12)}$ with 80-by-100 dimension and $R^{(23)}$ with 100-by-80 dimension, are binary matrices with 2-by-2 block structures. $R^{(12)}$ is generated based on the block structure $\left[\begin{smallmatrix} 0.9 & 0.7 \\ 0.8 & 0.9 \end{smallmatrix}\right]$, i.e., the objects in cluster 1 of $\mathcal{X}^{(1)}$ is related to the objects in cluster 1 of $\mathcal{X}^{(2)}$ with probability 0.9, and so on so forth. $R^{(23)}$ is generated based on the block structure $\left[\begin{smallmatrix} 0.6 & 0.7 \\ 0.7 & 0.6 \end{smallmatrix}\right]$. Each type of objects has two equal size clusters. It is not a trivial task to identify the cluster structure of this data, since the block structures are subtle. We denote this data as Binary Relation Matrices (TRM) data.

Other three data sets are built based on the 20-newsgroups data for hierarchical taxonomy mining and document clustering. In the field of text categorization, hierarchical taxonomy classification is widely used to obtain a better trade-off between effectiveness and efficiency than flat taxonomy classification. To take advantage of hierarchical classification, one must mine a hierarchical taxonomy from the data set. We can see that words, documents, and categories formulate a tri-type relational data, which consist of two relation matrices: a word-document matrix $R^{(12)}$ and a document-category matrix

TABLE 17.6: Taxonomy structures for three data sets

Data set	Taxonomy structure
TM1	{NG10, NG11}, {NG17, NG18, NG19}
TM2	{NG2, NG3}, {NG8, NG9}, {NG12, NG13}
TM3	{NG4, NG5}, {NG8, NG9}, {NG14, NG15}, {NG17, NG18}

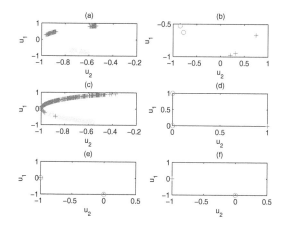

FIGURE 17.5: Three pairs of embeddings of documents and categories for the TM1 data set produced by SRC with different weights: (a) and (b) with $w_a^{(12)} = 1, w_a^{(23)} = 1$; (c) and (d) with $w_a^{(12)} = 1, w_a^{(23)} = 0$; (e) and (f) with $w_a^{(12)} = 0, w_a^{(23)} = 1$.

$R^{(23)}$ [56].

The true taxonomy structures for three data sets, TM1, TM2, and TM3, are listed in Table 17.6. For example, TM1 data set is sampled from five categories in which NG10 and NG11 belong to the same high-level category *res.sports* and NG17, NG18, and NG19 belong to the same high-level category *talk.politics*. Therefore, for the TM1 data set, the expected clustering result on categories should be {NG10, NG11} and {NG17, NG18, NG19} and the documents should be clustered into two clusters according to their categories. The documents in each data set are generated by sampling 100 documents from each category.

The number of clusters used for documents and categories are 2, 3, and 4 for TM1, TM2, and TM3, respectively. For the number of word clusters, we adopt the number of categories, i.e., 5, 6, and 8. For the weights $w_a^{(12)}$ and $w_a^{(23)}$, we simply use equal weight, i.e., $w_a^{(12)} = w_a^{(23)} = 1$. Figure 17.5 illustrates the effects of different weights on embeddings of documents and categories. When $w_a^{(12)} = w_a^{(23)} = 1$, i.e., SRC makes use of both word-document re-

TABLE 17.7:   NMI comparisons of SRC,
MRK, and CBGC algorithms

Data set	SRC	MRK	CBGC
BRM	0.6718	0.6470	0.4694
TM1	1	0.5243	–
TM2	0.7179	0.6277	–
TM3	0.6505	0.5719	–

lations and document-category relations, both documents and categories are separated into two clusters very well as in Figure 17.5a and b, respectively; when SRC makes use of only the word-document relations, the documents are separated with partial overlapping as in Figure 17.5c and the categories are randomly mapped to a couple of points as in Figure 17.5d; when SRC makes use of only the document-category relations, both documents and categories are incorrectly overlapped as in Figure 17.5e and f, respectively, since the document-category matrix itself does not provide any useful information for the taxonomy structure.

The performance comparison is based on the cluster quality of documents, since the better it is, the more accurate we can identify the taxonomy structures. Table 17.7 shows NMI comparisons of the three algorithms on the four data sets. The NMI score of CBGC is available only for BRM data set because the CBGC program provided by the authors only works for the case of two clusters and small-size matrices. We observe that SRC performs better than MRK and CBGC on all data sets. The comparison shows that among the limited efforts in the literature attempting to cluster multi-type interrelated objects simultaneously, SRC is an effective one to identify the cluster structures of MTRD.

Chapter 18

Multiple-View and Evolutionary Data Clustering

In this chapter, we apply our multiple-view algorithms to different applications with both synthetic data and real data. We show that by using multiple patterns from multiple-views, our algorithms provide more accurate and robust patterns on various data sets.

18.1 Multiple-View Clustering

In this section, we evaluate the effectiveness of Multiple-View Clustering (MVC) algorithm. For the comparisons, the average performance of all the views (call it AV) and the best performance of all the views (call it BV) are reported. As for other multiple-view clustering algorithms, there are limited efforts in the literature on multiple-view clustering [19,37,139], which focus on special cases with two views or a certain type of representation (Section 6.2). We use the classic algorithm proposed in [19] (call it Two-View Clustering (TVC)) as a comparison when it is applicable. For performance measure, we elect to use the Normalized Mutual Information (NMI) [119] between the resulting cluster labels and the true cluster labels, which is a standard way to measure the cluster quality. The final performance score is the average of 10 runs. For the number of clusters k, we simply use the number of true clusters. The k-means algorithm is used as the base clustering algorithm except stated otherwise. Random initializations are used for MVC. For the value of α in MVC, we simply adopt 1 in all experiments, since we find that the results in the experiments are not sensitive to the change of α.

18.1.1 Synthetic Data

The first synthetic data set (syn1) is a toy example consisting of three views for letters "SDM." The first view is in a two-dimension space such that "D" and "M" are close; the second view is also in a two-dimension space such that "S" and "D" are close; the third view is a graph in which the data points consisting of the letters are linked to each other with certain probabilities, 0.2

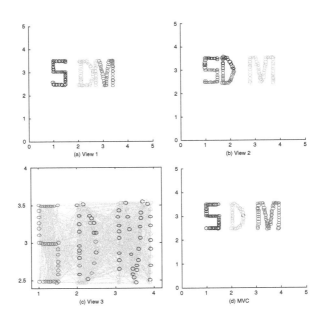

FIGURE 18.1: A toy example that demonstrates that our MVC algorithm is able to learn the consensus pattern from multiple-views with noise.

TABLE 18.1:   Distributions and parameters to generate syn2 data

	View 1	View 2	View 3	View 4
Cluster 1	N(1,1)	B(1)	P(2)	E(1)
Cluster 2	N(2,1)	B(0.2)	P(6)	E(25)
Cluster 3	N(5,1)	B(0.1)	P(12)	E(50)

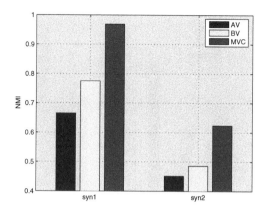

FIGURE 18.2: NMI comparison on synthetic data.

for the data points in the same letter and 0.1 for the data points in different letters. Figure 18.1a shows the clustering result of k-means algorithm on view 1, and we observe that "D" and "M" are confused. Figure 18.1b shows the clustering result of k-means algorithm on view 2, and we observe that "S" and "D" are confused. Figure 18.1c shows the clustering result of the graph clustering algorithm, METIS [75], on view 3, and we observe that three letters are confused. Figure 18.1(d) provides a visualization for the clustering pattern which is learned by MVC algorithm based on the three "confusing" patterns from the three views, and we observe that the three letters are almost perfectly distinguished from each other. The corresponding NMI scores of AV, BV, and MVC are shown in Figure 18.2.

The second synthetic data set (syn2) is to simulate a typical situation: different features from different views have totally different distributions. The distributions and parameters to generate syn2 data are summarized in Table 18.1 in which "N" denotes normal distribution; "P" denotes Poisson distribution; "B" denotes Bernoulli distribution; "E" denotes exponential distribution. The size of syn2 is 9000 and 3000 for each cluster. We use the appropriate EM algorithms to cluster each view, i.e., Gaussian EM, Bernoulli EM, Poisson EM, and Exponential EM [13] are for views 1, 2, 3, and 4, respectively. NMI comparison on syn2 data is shown in Figure 18.2. We observe that MVC's performances are significantly better than average and best performance based on each single view.

In summary, these experiments based on synthetic data sets demonstrate that the MVC algorithm can effectively learn the consensus clustering pattern, which is more accurate and robust than patterns based on any single view, from multiple representations with very different forms and different statistical properties.

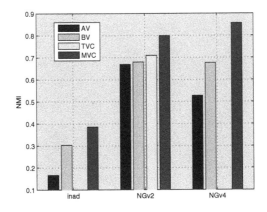

FIGURE 18.3: NMI comparison on real data.

18.1.2   Real Data

We evaluate the performance of MVC algorithm on three real data sets. The first data set is Internet Advertisement Data Set (INTAD) [81] in which each instance is an image in a Web page and belongs to either of two clusters, ad or non-ad. Each instance is described from six views: geometry of the image, phrases occurring in the URL, the image's URL, alt text, the anchor text, and words occurring near the anchor text. INTAD data set has 3279 instances (2821 non-ads and 458 ads) and totally 1558 features (continuous or binary) for six views. Multiple-views for INTAD data set are naturally formulated and we do not know if the views are independent or not. However, this issue does not affect MVC algorithm

The second and third data sets are based on the 20-newsgroups [?], which was used in [19] to test the multiple-view clustering algorithm, TVC. We follow the same way as in [19] to generate two multiple-view data sets by concatenating TF-IDF term vectors randomly drawn from different newsgroups (please refer to [19] for details). [19] uses this construction to ensure the data to satisfy the independent assumption. The cluster and view structures for two data sets, NGv2 and NGv4, are given as follows:

$$
\begin{bmatrix}
 & cluster1 & cluster2 \\
view1 & windows.misc & pc.hardware \\
view2 & politics.mideast & politics.misc
\end{bmatrix}
$$

$$
\begin{bmatrix}
 & cluster1 & cluster2 & cluster3 \\
view1 & alt.atheism & comp.graphics & windows.misc \\
view2 & windows.x & misc.forsale & rec.autos \\
view3 & sport.hockey & sci.crypt & sci.electronics \\
view4 & politics.guns & politics.mideast & politics.misc
\end{bmatrix}
$$

We use 2000 features for each view and 500 documents for each cluster. Hence, NGv2 has 1000 documents with 4000 features and NGv4 has 1500 documents with 8000 features.

The NMI comparison on real data sets is shown in Figure 18.3. We observe that MVC provides the best performance on all data sets. For the difficult INTAD data set, MVC's performance increases about 100% compared with the average performance based on all single views and 20% compared with the best performance of all single views. This shows that MVC is capable of learning the consensus pattern even from very noisy multiple patterns. MVC also performs significantly better than TVC on NGv2 data set (for other data set with more than two views, TVC is not applicable). We do not have definite reason for this, since the theory behind TVC algorithm is still not completely clear and actually TVC's convergence is not guaranteed.

In summary, above experiments show that no matter whether the data satisfy the independence assumption, MVC algorithm always efficiently learns robust consensus patterns from multiple-view data with different properties and different levels of noise.

18.2   Multiple-View Spectral Embedding

To test the effectiveness of Multiple-View Spectral Embedding (MVSE) algorithm on high-dimensional data, we elect to use high-dimensional 20-newsgroups data. Similarly, we construct NGv3 data set with the following cluster and view structures:

$$
\begin{bmatrix}
 & cluster1 & cluster2 \\
view1 & windows.misc & pc.hardware \\
view2 & sport.baseball & sport.hockey \\
view3 & politics.mideast & politics.misc
\end{bmatrix}
$$

Hence, NGv3 is a high-dimensional data set with 6000 dimensions. We reduce it to two-dimensional spaces. Figure 18.4 shows four embeddings for NGv3 data set. In Figure 18.4, each embedding is in a two-dimension eigen-space corresponding to the two leading eigenvectors. Figure 18.4a, b, and c shows the embeddings for views 1, 2, and 3, respectively. Figure 18.4d shows the embedding provided by the MVSE algorithm. We observe that MVSE provides an embedding with the best separation, i.e., the intrinsic cluster patterns of the data are clearer in MVSE embedding than in the embeddings from the three single views.

Since direct observation is not accurate, we use the following method to evaluate which embedding has less noise and provides clearer pattern. We run k-means on each embedding with exactly the same initialization and then compare the clustering results from different embeddings by NMI scores. We

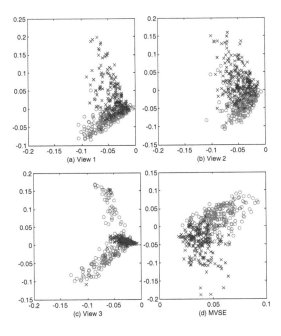

FIGURE 18.4: Four embeddings for NGv3 data set.

expect that the clearer patterns an embedding has, the better the clustering result is. The NMI comparison is shown in Figure 18.5. We observe that the NMI score for MVSE embedding shows about 200% improvement even compared with the best performance for the single-view embeddings. We also conduct experiments on NGv4 data set with 8000 dimensions. We reduce it to three-dimensional eigen-spaces. The NMI scores are also shown in Figure 18.5. We observe the similar result to that for NGv3.

In summary, as the first algorithm to address multiple-view dimensionality reduction, MVSE demonstrates its effectiveness by providing optimal spectral embeddings, which is more robust to noise and has a better separation than the spectral embeddings based on a single view.

18.3   Semi-Supervised Clustering

As we discussed in Section 12.2, our model and algorithms can be extended to deal with unsupervised clustering with side information. In this section, we use MVC to perform semi-supervised clustering on INTAD data set to

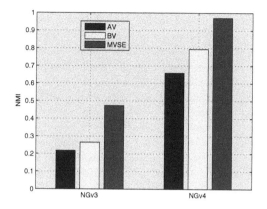

FIGURE 18.5: NMI comparison on spectral embedding of NGv3 and NGv4.

demonstrate the great potential of this extension.

We randomly choose a certain percent of instances from INTAD data set and label them as side information. Then we run MVC on the partially labeled INTAD data set in comparison with a state-of-the-art semi-supervised clustering algorithm, HMRF-KMEANS algorithm [14].

Figure 18.6 shows the clustering results on INTAD data set with different percents of instances labeled. We observe that both MVC and HMRF-KMEANS's performances increase with the number of labeled instances. MVC's performance is comparable with the classic HMRF-KMEANS algorithm. This demonstrates the great potential of MVC on semi-supervised clustering, since MVC is flexible to adopt different types of clustering algorithm instead of only k-means algorithm.

18.4 Evolutionary Clustering

As a fairly new filed, there are short of benchmark data with ground truth for evolutionary clustering and performance measures for evolutionary clustering are under development [27]. In this section, we use a synthetic data set with true labels available to show the great potential of evolutionary MVC, which provides more accurate clustering patterns at each time step by incorporating temporal smoothness.

We construct a mixture of Gaussian distributions evolving with time. At the start time, three components of 6000 data points have means 1, 4, 7 and the same variance 1. At each time t, the means of the three cluster centers move

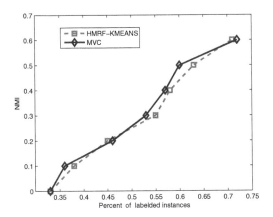

FIGURE 18.6: Semi-supervised clustering results.

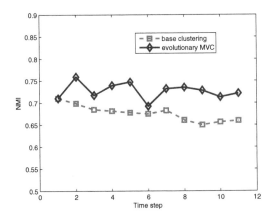

FIGURE 18.7: Evolutionary clustering results.

a small distance with distribution $N(0, 0.1)$ and the variance increases by 0.1. Hence, the cluster centers are drifting and expanding over time. Figure 18.7 shows the clustering results of the evolutionary MVC and the base clustering algorithm, k-means. We observe that MVC performs significantly better than the base algorithm at most time steps. We also observe that the base clustering patterns degenerate over time, and, on the other hand, the clustering patterns from evolutionary MVC do not show a clear degenerating trend. In fact, the clustering patten for evolutionary MVC at last time step is still significantly better than the start pattern. This is because by efficiently combining the patterns at time $t - 1$ and time t, evolutionary MVC is more robust to noise at each time step and, by incorporating temporal smoothness, the intrinsic pattern of the data is passed on over time by evolutionary MVC to avoid fast degeneration of the patterns. For the evolutionary clustering applications by DPChain, HDP-EVO, and HDP-HTM models, please refer to [131, 132].

Part IV

Summary

In this book, we introduce a novel theoretical framework for a new data mining field, relational data clustering, and a family of new algorithms for different relational clustering problems arising in a wide range of important applications. Specifically, we cover the model, algorithm, and application aspects for six basic relational clustering problems: co-clustering co-clustering, multi-type heterogeneous relational data clustering, homogeneous relational data clustering, general relational data clustering, multiple-view relational data clustering, and evolutionary data clustering.

Co-clustering focuses on bi-type heterogeneous relational data in which there are heterogeneous relations between the two types of data objects. For example, a text corpus can be formulated as a bi-type relational data set of documents and words in which there exist heterogeneous relations between documents and words. We propose a new co-clustering framework called Block Value Decomposition (BVD). The key idea is that the latent block structure in a two-dimensional dyadic data matrix can be explored by its triple decomposition. The dyadic data matrix is factorized into three components: the row-coefficient matrix R, the block value matrix B, and the column-coefficient matrix C. The coefficients denote the degrees of the rows and columns associated with their clusters and the block value matrix is an explicit and compact representation of the hidden block structure of the data matrix. Under this framework, we develop a specific novel co-clustering algorithm for a special yet very popular case—nonnegative dyadic data, that iteratively computes the three decomposition matrices based on the multiplicative updating rules derived from an objective criterion. By intertwining the row clusterings and the column clusterings at each iteration, the algorithm performs an implicitly adaptive dimensionality reduction, which works well for typical high-dimensional and sparse data in many data mining applications. We have proven the correctness of the algorithm by showing that the algorithm is guaranteed to converge.

Multi-type heterogeneous relational data clustering is more general than bi-type heterogeneous relational data, which can be formulated as k-partite graphs with various structures. In fact, many examples of real-world data involve multiple types of data objects that are related to each other, which naturally form k-partite graphs of heterogeneous types of data objects. For example, documents, words, and categories in taxonomy mining, as well as Web pages, search queries, and Web users in a Web search system all form a tripartite graph; papers, key words, authors, and publication venues in a scientific publication archive form a quart-partite graph. To address heterogeneous relational data clustering problem, we propose a general model, the relation summary network, to find the hidden structures (the local cluster structures and the global community structures) from a k-partite graph. The basic idea is to construct a new k-partite graph with hidden nodes, which "summarize" the link information in the original k-partite graph and make the hidden structures explicit, to approximate the original graph. The model provides a principal framework for unsupervised learning on k-partite graphs of various structures. Second, under this model, based on the matrix representation of a k-partite graph we reformulate the graph approximation as an optimization problem of matrix approximation and derive an iterative algorithm to find the hidden structures from a k-partite graph under a broad range of distortion measures. By iteratively updating the cluster structures for each type of nodes, the algorithm takes advantage of the interactions among the cluster structures of different types of nodes and performs implicit adaptive feature reduction for each type of nodes. Third, we also establish the connections between existing clustering approaches and the proposed model to provide a unified view to the clustering approaches.

Another important relational clustering problem is homogeneous relational data clustering. In heterogeneous relational data, we have heterogeneous relations between different types of data objects. On the other hand, in homogeneous relational data, there are homogeneous relations between the data objects of a single type. Homogeneous relational data also arise from important applications, such as Web mining, social network analysis, bioinformatics, VLSI design, and task scheduling. Graph partitioning in the literature can be viewed as a special case of homogeneous relational data clustering. Basically, graph partitioning looks for dense clusters corresponding to strongly intra-connected subgraphs. On the other hand, the goal of homogeneous relational data clustering is more general and challenging. It is to identify both dense clusters and sparse clusters. we propose a general model based on graph approximation to learn the link-pattern-based cluster structure from a graph. By unifying the traditional edge cut objectives, the model provides a new view to understand graph partitioning approaches and at the same time it is applicable to learning various cluster structures. Under this model, we derive three novel algorithms to learn the general cluster structures from a graph, which cover three main versions of unsupervised learning algorithms, hard, soft, and balanced versons, to provide a complete family of cluster learning algorithms. This family of algorithms has the following advantages: they are flexible to learn various types of clusters; when applied to learning strongly intra-connected clusters, this family evolves to a new family of effective graph partition algorithms; it is easy for the proposed algorithms to incorporate the prior knowledge of the cluster structure into the algorithms.

The most general case of relational data contain three types of information: attributes for individual objects, homogeneous relations between objects of the same type, and heterogeneous relations between objects of different types. How to make use of all three types of information to cluster multi-type-related objects simultaneously is a big challenge, since the three types of information have different forms and very different statistical properties. we propose a general probabilistic model for relational clustering, mixed membership relational data clustering, which also provides a principal framework to unify various important clustering tasks, including traditional attributes-based clustering, semi-supervised clustering, co-clustering, and graph clustering. The proposed model seeks to identify cluster structures for each type of data objects and interaction patterns between different types of objects. It is applicable to relational data of various structures. Under this model, we propose parametric hard and soft relational clustering algorithms under a large number of exponential family distributions. The algorithms are applicable to various relational data from various applications and at the same time unify a number of stat-of-the-art clustering algorithms: co-clustering algorithms, the k-partite graph clustering, Bregman k-means, and semi-supervised clustering based on hidden Markov random fields.

Then, we address relational data clustering problems under individual relational clustering framework. We propose a general model for multiple-view unsupervised learning. The proposed model introduces the concept of mapping function to make the different patterns from different pattern spaces comparable and hence an optimal pattern can be learned from the multiple patterns of multiple representations. We propose a general model for multiple-view unsupervised learning, which is applicable to various types of unsupervised learning on various types of multiple-view data. We show how to formulate the problems of multiple-view clustering and multiple view spectral embedding (dimensionality reduction). We derive an iterating algorithm to solve the constrained non-convex problem of multiple view clustering, which iteratively updates the optimal clustering pattern matrix and the mapping matrices until convergence. We prove the objective function is nonincreasing under the updating rules and hence the convergence of the algorithm is guaranteed. We derive a simple algorithm for multiple spectral embedding, which provides a global optimal solution to the problem. We also introduce extensions to our algorithms to handel two important learning settings: unsupervised learning with side information and evolutionary clustering.

Last, we present our most recent research on the evolutionary clustering, which has great potential to incorporate time effects into relational data clustering. Evolutionary clustering is a relatively new research topic in data mining. Evolutionary clustering refers to the scenario where a collection of data evolves over the time; at each time, the collection of the data has a number of clusters; when the collection of the data evolves from one time to another, new data items may join the collection and existing data items may disappear; similarly, new clusters may appear and at the same time existing clusters may disappear. Consequently, both the data items and the clusters of the collection may change over the time, which poses a great challenge to the problem of evolutionary clustering in comparison with the traditional clustering. We propose a statistical approach to solving the evolutionary clustering problem. We assume that the cluster structure at each time follows a mixture model of the clusters for the data collection at this time; clusters at different times may share common clusters; further, these clusters evolve over the time and some may become more popular while others may become outdated, making the cluster structures and the number of clusters change over the time. Consequently, we use Dirichlet Process (DP) to model the evolutionary change of the clusters over the time. Specifically, we propose three Dirichlet process-based models for the evolutionary clustering problem: DPChain, HDP-EVO, HDP-HTM.

References

1. N. M. Laird, A. P. Dempster, and D. B. Rubin. Maximum likelihood from incomplete data via the EM algorithm. *Journal of the Royal Statistical Society*, 39(8):1–38, 1977.

2. Steven Abney. Bootstrapping. In *ACL '02*, pages 360–367, 2002.

3. Charu C. Aggarwal, Cecilia Magdalena Procopiuc, Joel L. Wolf, Philip S. Yu, and Jong Soo Park. Fast algorithms for projected clustering. In *SIGMOD Conference*, pages 61–72, 1999.

4. E. M. Airoldi, D. M. Blei, E. P. Xing, and S. E. Fienberg. Mixed membership stochastic block models for relational data with application to protein-protein interactions. In *ENAR-2006*.

5. A. K. Jain and R. C. Dubes. *Algorithms for Clustering Data*. Prentice-Hall, Englewood Cliffs, 1988.

6. David Aldous. Exchangeability and related topics. *Ecole de Probabilites de Saint-Flour*, (XIII):1–198, 1983.

7. C. Antoniak. Mixtures of Dirichlet processes with applications to bayesian nonparametric problems. *The Annals of Statistics*, 2(6):1152–1174, 1974.

8. Francis R. Bach and Michael I. Jordan. Learning spectral clustering. In Sebastian Thrun, Lawrence Saul, and Bernhard Schölkopf, editors, *Advances in Neural Information Processing Systems 16*, 2004.

9. A. Banerjee, I. Dhillon, J. Ghosh, and S. Sra. Generative model-based clustering of directional data. In *KDD'03*, 2003.

10. Arindam Banerjee and Inderjit Dhillon. Generative model-based clustering of directional data. In *In Proceedings of the Ninth ACM SIGKDD International Conference on Knowledge Discovery and Data Mining (KDD-2003)*, 2003.

11. Arindam Banerjee, Inderjit S. Dhillon, Joydeep Ghosh, Srujana Merugu, and Dharmendra S. Modha. A generalized maximum entropy approach to Bregman co-clustering and matrix approximation. In *KDD*, pages 509–514, 2004.

12. Arindam Banerjee, Srujana Merugu, Inderjit S. Dhillon, and Joydeep Ghosh. Clustering with Bregman divergences. In *SDM*, 2004.

13. Arindam Banerjee, Srujana Merugu, Inderjit S. Dhillon, and Joydeep Ghosh. Clustering with Bregman divergences. *Journal of Machine Learning Research*, 6:1705–1749, 2005.

14. Sugato Basu, Mikhail Bilenko, and Raymond J. Mooney. A probabilistic framework for semi-supervised clustering, In *KDD '04: Proceedings of the tenth ACM SIGKDD international conference on Knowledge discovery and data mining* pages 59–68, Seattle, WA, August 2004.

15. Matthew J. Beal, Zoubin Ghahramani, and Carl Edward Rasmussen. The infinite hidden Markov model. In *NIPS 14*, 2002.

16. Pavel Berkhin. Survey of clustering data mining techniques. *Technical Report*, Accrue Software, San Jose, CA, 2002.

17. R. Bhatia. *Matrix Analysis*. Springer-Verlag, New York, 1997.

18. I. Bhattachrya and L. Getor. Entity resolution in graph data. *Technical Report CS-TR-4758*, University of Maryland, 2005.

19. Steffen Bickel and Tobias Scheffer. Multi-view clustering. In *ICDM '04*, pages 19–26, 2004.

20. D. Blackwell and J. MacQueen. Ferguson distributions via polya urn schemes. *The Annals of Statistics*, 1(2):353–355, 1973.

21. D. Blei, A. Ng, and M. Jordan. Latent Dirichlet allocation. *Journal of Machine Learning Research*, 3:993–1022, January 2003.

22. David Blei and John Lafferty. Dynamic topic models. In *In Proceedings of the 23rd International Conference on Machine Learning*, 2006.

23. Avrim Blum and Tom Mitchell. Combining labeled and unlabeled data with co-training. In *COLT' 98*, pages 92–100, 1998.

24. Ulf Brefeld and Tobias Scheffer. Co-em support vector learning. In *ICML '04*, page 16, 2004.

25. Thang Nguyen Bui and Curt Jones. A heuristic for reducing fill-in in sparse matrix factorization. In *PPSC*, pages 445–452, 1993.

26. George Casella and Edward I. George. Explaining the Gibbs sampler. *The American Statistician*, 46(3):167–174, August 1992.

27. Deepayan Chakrabarti, Ravi Kumar, and Andrew Tomkins. Evolutionary clustering. In *Proceedings of the 12th ACM SIGKDD International Conference on Knowledge Discovery and Data Mining*, pages 554–560, 2006.

28. Pak K. Chan, Martine D. F. Schlag, and Jason Y. Zien. Spectral k-way ratio-cut partitioning and clustering. In *DAC '93*, pages 749–754, 1993.

29. Yizong Cheng and George M. Church. Biclustering of expression data. In *ICMB*, pages 93–103, 2000.

30. Yun Chi, Xiaodan Song, Dengyong Zhou, Koji Hino, and Belle L. Tseng. Evolutionary spectral clustering by incorporating temporal smoothness. In *Proceedings of the 13th ACM SIGKDD International Conference on Knowledge Discovery and Data Mining*, pages 153–162, 2007.

31. H. Cho, I. Dhillon, Y. Guan, and S. Sra. Minimum sum squared residue co-clustering of gene expression data. In *SDM*, 2004.

32. Aaron Clauset, M. E. J. Newman, and Christopher Moore. Finding community structure in very large networks. *Phys. Rev. E*, 70(6):66–111, 2004.

33. M. Collins, S. Dasgupta, and R. Reina. A generalizaionof principal component analysis to the exponential family. In *NIPS'01*, 2001.

34. M. Collins and Y. Singer. Unsupervised models for named entity classification, In *Proceedings of the Joint SIGDAT Conference on Empirical Methods in Natural Language Processing and Very Large Corpora*, pages 100–110, 1999.

35. Sanjoy Dasgupta, Michael L. Littman, and David A. McAllester. Pac generalization bounds for co-training. In *NIPS*, 2001.

36. D. D. Lee and H. S. Seung. Learning the parts of objects by non-negative matrix factorization. *Nature*, 401:788–791, 1999.

37. Virginia R. de Sa. Spectral clustering with two views. In *ICML Workshop on Learning with Multiple Views*, 2005.

38. Scott C. Deerwester, Susan T. Dumais, Thomas K. Landauer, George W. Furnas, and Richard A. Harshman. Indexing by latent semantic analysis. *Journal of the American Society of Information Science*, 41(6):391–407, 1990.

39. Arthur Dempster, Nan Laird, and Donald Rubin. Maximum likelihood from incomplete data via the em algorithm. *Journal of the Royal Statistical Society*, 39(1):1–38, 1977.

40. Inderjit Dhillon, Yuqiang Guan, and Brian Kulis. A unified view of kernel k-means, spectral clustering and graph cuts. *Technical Report TR-04-25*, University of Texas at Austin, 2004.

41. Inderjit Dhillon, Yuqiang Guan, and Brian Kulis. A fast kernel-based multilevel algorithm for graph clustering. In *KDD '05*, 2005.

42. Inderjit Dhillon, Yuqiang Guan, and Brian Kulis. A fast kernel-based multilevel algorithm for graph clustering. In *KDD '05*, pages 629–634, 2005.

43. Inderjit S. Dhillon. Co-clustering documents and words using bipartite spectral graph partitioning. In *KDD*, pages 269–274, 2001.

44. Inderjit S. Dhillon, Subramanyam Mallela, and Dharmendra S. Modha. Information-theoretic co-clustering. In *KDD '03: Proceedings of the ninth ACM SIGKDD International Conference on Knowledge Discovery and Data Mining*, pages 89–98, New York, NY, 2003. ACM.

45. Chris Ding, Xiaofei He, and H. D. Simon. On the equivalence of non-negative matrix factorization and spectral clustering. In *SDM'05*, 2005.

46. Chris H. Q. Ding and Xiaofeng He. Linearized cluster assignment via spectral ordering. In *ICML*, 2004.

47. Chris H. Q. Ding, Xiaofeng He, Hongyuan Zha, Ming Gu, and Horst D. Simon. A min-max cut algorithm for graph partitioning and data clustering. In *Proceedings of ICDM 2001*, pages 107–114, 2001.

48. Saso Dzeroski and Nada Lavrac, editors. *Relational Data Mining*. Springer-Verlag, NY, 2000.

49. Ran El-Yaniv and Oren Souroujon. Iterative double clustering for unsupervised and semi-supervised learning. In *ECML*, pages 121–132, 2001.

50. E. Erosheva and S. E. Fienberg. Bayesian mixed membership models for soft clustering and classification. *Classification: The Ubiquitous Challenge*, Springer, Heidelberg, pages 11–26, 2005.

51. E. A. Erosheva, S. E. Fienberg, and J. Lafferty. Mixed membership models of scientific publications. In *NAS*.

52. Michael D. Escobar and Mike West. Bayesian density estimation and inference using mixtures. *The Annals of Statistics*, 90:577–588, 1995.

53. Thomas S Ferguson. A bayesian analysis of some nonparametric problems. *The Annals of Statistics*, 1(2):209–230, 1973.

54. Xiaoli Zhang Fern and Carla E. Brodley. Solving cluster ensemble problems by bipartite graph partitioning. In *ICML '04*, 2004.

55. S. E. Fienberg, M. M. Meyer, and S. Wasserman. Satistical analysis of multiple cociometric relations. *Journal of American Satistical Association*, 80:51–87, 1985.

56. Bin Gao, Tie-Yan Liu, Xin Zheng, Qian-Sheng Cheng, and Wei-Ying Ma. Consistent bipartite graph co-partitioning for star-structured high-order heterogeneous data co-clustering. In *KDD '05*, pages 41–50, 2005.

57. Lise Getoor. An introduction to probabilistic graphical models for relational data. *Data Engineering Bulletin*, 29, 2006.

58. G. Golub and C. Van Loan. *Matrix Computations*. Johns Hopkins University Press, Baltimore, MD, 1989.

59. Rayid Ghani. Combining labeled and unlabeled data for text classification with a large number of categories. In *ICDM '01*, pages 597–598, 2001.

60. Michelle Girvan and M. E. J. Newman. Community structure in social and biological networks. In *National Academic Science*, 2001.

61. Patrick Glenisson, Janick Mathys, and Bart De Moor. Meta-clustering of gene expression data and literature-based information. *ACM SIGKDD Explorations Newsletter*, 5(2):101–112, 2003.

62. Thomas L. Griffiths and Mark Steyvers. Finding scientific topics. In *Proceedings of the National Academy of Sciences*, pages 5228–5235, February, 2004.

63. Bruce Hendrickson and Robert Leland. A multilevel algorithm for partitioning graphs. In *Supercomputing '95: Proceedings of the 1995 ACM/IEEE conference on Supercomputing (CDROM)*, page 28, 1995.

64. M. Henzinger, R. Motwani, and C. Silverstein. Challenges in web search engines. In *Proceedings of the 18th International Joint Conference on Artificial Intelligence*, pages 1573–1579, 2003.

65. P. D. Hoff, A. E. Rafery, and M. S. Handcock. Latent space approaches to social network analysis. *Journal of American Satistical Association*, 97:1090–1098, 2002.

66. Thomas Hofmann. Probabilistic latent semantic analysis. In *Proceedings of Uncertainty in Artificial Intelligence, UAI'99*, Stockholm, 1999.

67. Thomas Hofmann and Jan Puzicha. Latent class models for collaborative filtering. In *IJCAI'99*, Stockholm, 1999.

68. Lawrence B. Holder and Diane J. Cook. Graph-based relational learning: current and future directions. *ACM SIGKDD Explorations Newsletter*, 5(1):90–93, 2003.

69. M. Gu X. he H. Zha, C. Ding, and H. Simon. Bi-partite graph partitioning and data clustering. In *ACM CIKM'01*, 2001.

70. Hemant Ishwaran and Lancelot F. JAMES. Gibbs sampling methods for stick-breaking priors. *Journal of the American Statistical Association*, 96(453):161–173, 2001.

71. J. A. Hartigan. Direct clustering of a data matrix. *Journal of the American Statistical Association*, 67(337):123–129, March 1972.

72. A. K. Jain, M. N. Murty, and P. J. Flynn. Data clustering: a review. *ACM Computing Surveys*, 31(3):264–323, 1999.

73. G. Jeh and J. Widom. Simrank: A measure of structural-context similarity. In *KDD-2002*, 2002.

74. Hiroyuki Kaji and Yasutsugu Morimoto. Unsupervised word sense disambiguation using bilingual comparable corpora. In *Proceedings of the 19th COLIN*, pages 1–7, 2002.

75. George Karypis and Vipin Kumar. A fast and high quality multilevel scheme for partitioning irregular graphs. *SIAM Journal on Scientific Computing*, 20(1):359–392, 1998.

76. M. Kearns, Y. Mansour, and A. Ng. An information-theoretic analysis of hard and soft assignment methods for clustering. In *UAI'97*, pages 282–293, 2004.

77. B. Kernighan and S. Lin. An efficient heuristic procedure for partitioning graphs. *The Bell System Technical Journal*, 49(2):291–307, 1970.

78. Mathias Kirsten and Stefan Wrobel. Relational distance-based clustering. In *Proceedings of Fachgruppentreffen Maschinelles Lernen (FGML-98)*, pages 119–124, 1998.

79. Jon M. Kleinberg. Authoritative sources in a hyperlinked environment. *Journal of the ACM*, 46(5):604–632, 1999.

80. Ravi Kumar, Prabhakar Raghavan, Sridhar Rajagopalan, and Andrew Tomkins. Trawling the Web for emerging cyber-communities. *Computer Networks (Amsterdam, Netherlands: 1999)*, 31(11–16):1481–1493, 1999.

81. Nicholas Kushmerick. Learning to remove internet advertisements. In *AGENTS '99*, pages 175–181, 1999.

82. Ken Lang. NewsWeeder: learning to filter netnews. In *ICML'95*, pages 331–339, 1995.

83. Daniel D. Lee and H. Sebastian Seung. Algorithms for non-negative matrix factorization. In *NIPS*, pages 556–562, 2000.

84. Honglak Lee, Alexis Battle, Rajat Raina, and Andrew Y. Ng. Efficient sparse coding algorithms. In *NIPS'07*, pages 801–808, 2007.

85. Tao Li. A general model for clustering binary data. In *KDD'05*, 2005.

86. Bo Long, Xiaoyun Wu, Zhongfei (Mark) Zhang, and Philip S. Yu. Unsupervised learning on k-partite graphs. In *KDD-2006*, 2006.

87. Bo Long, Zhongfei (Mark) Zhang, Xiaoyun Wu, and Philip S. Yu. Spectral clustering for multi-type relational data. In *ICML'06*, 2006.

88. Bo Long, Zhongfei (Mark) Zhang, and Philip S. Yu. Co-clustering by block value decomposition. In *KDD'05*, 2005.

89. Bo Long, Zhongfei (Mark) Zhang, and Philip S. Yu. Combining multiple clusterings by soft correspondence. In *ICDM '05*, pages 282–289, 2005.

90. David J. C. MacKay. Ensemble learning for hidden Markov models. *Technical Report*, 1997.

91. K. V. Mardian. Statistics of directional data. *Journals of the Royal Statistical Society. Series B*, 37(3):349–393, 1975.

92. Radford M. Neal. Probabilistic inference using Markov chain monte carlo methods. *Technical Report*, (CRG-TR-93-1), 1993.

93. Radford M. Neal. Markov chain sampling methods for Dirichlet process mixture models. *Journal of Computational and Graphical Statistics*, 9(2):249–265, June 2000.

94. M. E. J. Newman. Detecting community structure in networks. *Physical Review*, 38:321–330, 2003.

95. M. E. J. Newman. Fast algorithm for detecting community structure in networks. *Physical Review*, 2004.

96. A. Ng, M. Jordan, and Y. Weiss. On spectral clustering: analysis and an algorithm. In *Advances in Neural Information Processing Systems 14*, 2001.

97. Andrew Y. Ng, Michael I. Jordan, and Yair Weiss. On spectral clustering: analysis and an algorithm. In *NIPS 14*, 2002.

98. J. C. Niebles, H. Wang, and L. Fei-Fei. Unsupervised activity perception by hierarchical bayesian models. In *British Machine Vision Conference (BMVC)*, 2006.

99. Kamal Nigam and Rayid Ghani. Analyzing the effectiveness and applicability of co-training. In *CIKM '00*, pages 86–93, 2000.

100. Kamal Nigam, Andrew K. McCallum, Sebastian Thrun, and Tom M. Mitchell. Text classification from labeled and unlabeled documents using EM. *Machine Learning*, 39(2/3):103–134, 2000.

101. L. R. Rabiner. A tutorial on hidden Markov models and selected applications inspeech recognition. In *Proceedings of the IEEE*, pages 257–286, 1989.

102. L. De Raedt and H. Blockeel. Using logical decision trees for clustering. In *Proceedings of the 7th International Workshop on Inductive Logic Programming*, 1997.

103. Stefan Rüping and Tobias Scheffer. Learning with multiple views. In *ICML Workshop on Learning with Multiple Views*, 2005.

104. P. Krishna Reddy and Masaru Kitsuregawa. Inferring web communities through relaxed cocitation and dense bipartite graphs. In *Data Basis Engineering Workshop*, 2001.

105. R. O. Duda, P. E. Hart, and D. G. Stork. *Pattern Classification.* John Wiley & Sons, New York, 2000.

106. N. A. Rosenberg, J. K. Pritchard, J. L. Weber, and H. M. Cann. Genetic structure of human population. *Science*, 298, 2002.

107. R. Salakhutdinov and S. Roweis. Adaptive overrelaxed bound optimization methods. In *ICML'03*, 2003.

108. Lawrence Saul and Fernando Pereira. Aggregate and mixed-order Markov models for statistical language processing. In *Proceedings of the Second Conference on Empirical Methods in Natural Language Processing*, 1997.

109. John P. Scott. *Social Network Analysis: A Handbook.* SAGE Publications, London, January 2000.

110. J. lafferty S. D. Pietra, V. D. Pietera. Duality and auxiliary functions for Bregman distances. *Technical Report CMU-CS-01-109*, Carnegie Mellon University, 2001.

111. Jayaram Sethuraman. A constructive definition of dirichlet priors. *Statistica Sinica*, 4:639–650, 1994.

112. S. Geman and D. Geman. Stochastic relaxation, Gibbs distribution, and the bayesian restoration of images. *Pattern Analysis and Machine Intelligence*, 6:721–742, 1984.

113. Jianbo Shi and Jitendra Malik. Normalized cuts and image segmentation. *IEEE Transactions on Pattern Analysis and Machine Intelligence*, 22(8):888–905, 2000.

114. Jianbo Shi and Jitendra Malik. Normalized cuts and image segmentation. *IEEE Transactions on Pattern Analysis and Machine Intelligence*, 22(8), August 2000.

115. Ajit Paul Singh and Geoffrey J. Gordon. Relational learning via collective matrix factorization. In *KDD*, pages 650–658, 2008.

116. Ajit Paul Singh and Geoffrey J. Gordon. A unified view of matrix factorization models. In *ECML/PKDD (2)*, pages 358–373, 2008.

117. Noam Slonim and Naftali Tishby. Document clustering using word clusters via the information bottleneck method. In *SIGIR '00*, 2000.

118. T. A. B. Snijders. Markov chain monte carlo estimation of exponential random graph models. *Journal of Social Structure*, 3:1-40, 2002.

119. Alexander Strehl and Joydeep Ghosh. Cluster ensembles — a knowledge reuse framework for combining partitionings. In *AAAI 2002*, pages 93–98, 2002.

120. Benjamin Taskar, Eran Segal, and Daphne Koller. Probabilistic classification and clustering in relational data. In *Proceeding of IJCAI-01*, 2001.

121. Y. Teh, M. Beal M. Jordan, and D. Blei. Hierarchical Dirichlet processes. *Journal of the American Statistical Association*, 101(476):1566–1581, 2007.

122. N. Tishby, F. Pereira, and W. Bialek. The information bottleneck method. In *Proceedings of the 37-th Annual Allerton Conference on Communication, Control and Computing*, pages 368–377, 1999.

123. Alexander Topchy, Anil K. Jain, and William Punch. Combining multiple weak clusterings. In *Proceedings of the Third IEEE International Conference on Data Mining*, page 331, 2003.

124. K. Wagstaff, C. Cardie, S. Rogers, and S. Schroedl. Constrained k-means clustering with background knowledge. In *ICML-2001*, pages 577–584, 2001.

125. Jidong Wang, Huajun Zeng, Zheng Chen, Hongjun Lu, Li Tao, and Wei-Ying Ma. Recom: reinforcement clustering of multi-type interrelated data objects. In *SIGIR '03*, pages 274–281, 2003.

126. X. Wang, X. Ma, and E. Grimson. Unsupervised activity perception by hierarchical bayesian models. In *Proceedings of IEEE Computer Society Conference on Computer Vision and Patter Recognition (CVPR)*, 2007.

127. Xuerui Wang and Andrew McCallum. Topics over time: a non-Markov continuous-time model of topical trends. In *Proceedings of the 12th ACM SIGKDD International Conference on Knowledge Discovery and Data Mining*, pages 424–433, 2006.

128. Andrew Y. Wu, Michael Garland, and Jiawei Han. Mining scale-free networks using geodesic clustering. In *KDD '04*, pages 719–724. ACM Press, 2004.

129. E. P. Xing, A. Y. Ng, M. I. Jorda, and S. Russel. Distance metric learning with applications to clustering with side information. In *NIPS'03*, volume 16, 2003.

130. Tianbing Xu, Zhongfei Zhang, Philip Yu, and Bo Long. Evolutionary clustering by hierarchical Dirichlet process with hidden Markov state. In *ICDM*, 2008.

131. Tianbing Xu, Zhongfei (Mark) Zhang, Philip S. Yu, and Bo Long. Dirichlet process based evolutionary clustering. In *ICDM '08: Proceedings of the 2008 Eighth IEEE International Conference on Data Mining*, pages 648–657, Washington, DC, 2008. IEEE Computer Society.

132. Tianbing Xu, Zhongfei (Mark) Zhang, Philip S. Yu, and Bo Long. Evolutionary clustering by hierarchical Dirichlet process with hidden Markov state. In *ICDM '08: Proceedings of the 2008 Eighth IEEE International Conference on Data Mining*, pages 658–667, Washington, DC, 2008. IEEE Computer Society.

133. Wei Xu, Xin Liu, and Yihong Gong. Document clustering based on non-negative matrix factorization. In *SIGIR '03*, pages 267–273, ACM Press, 2003.

134. X. Yin, J. Han, and P.S. Yu. Cross-relational clustering with user's guidance. In *KDD-2005*, 2005.

135. X. Yin, J. Han, and P.S. Yu. Linkclus: efficient clustering via heterogeneous semantic links. In *VLDB-2006*, 2006.

136. S. Yu and J. Shi. Multiclass spectral clustering. In *ICCV'03.*, 2003.

137. Hua-Jun Zeng, Zheng Chen, and Wei-Ying Ma. A unified framework for clustering heterogeneous web objects. In *WISE '02*, pages 161–172, 2002.

138. H. Zha, C. Ding, M. Gu, X. He, and H. Simon. Spectral relaxation for k-means clustering. *Advances in Neural Information Processing Systems*, 14, 2002.

139. Dengyong Zhou and Christopher J. C. Burges. Spectral clustering and transductive learning with multiple views. In *ICML '07*, pages 1159–1166, 2007.

140. Xiaojin Zhu, Zoubin Ghahramani, and John Lafferty. Time-sensitive Dirichlet process mixture models. *Technical Report*, (CMU-CALD-05-104), May 2005.

Index

auxiliary function, 78

Balanced CLGA, 103
bi-type heterogeneous relational data, 4, 181
bi-type heterogeneous relational data set, 11
Block Value Decomposition, 11, 14, 75
block value matrix, 16
block values, 16
Bregman divergences, 85, 106
BSGP, 90
BVD, 11, 75

CBGC, 147
CLGA, 35
clustering pattern matrix, 54
Co-clustering, 2, 3, 75, 119
collective clustering framework, 2
column-coefficient matrix, 12, 16

DP, 58
DPchain, 58
dyadic data, 13

EF-MMRC, 111
EM, 107
Evolutionary Clustering, 129
exponential families, 106
Exponential family MMRC, 111

general relational data clustering, 2
generalized I-divergence, 85
GGP, 36
Gibbs sampler, 108
Graph Clustering, 3, 120

Hard CLGA, 96

Hard MMRC Algorithm, 112
HDP, 57
HDP-EVO, 58
HDP-HTM, 58
heterogeneous relational data clustering, 2
heterogeneous relations, 2, 39
hidden block structure, 14
HMM, 57
HMRFs, 118
homogeneous relational data clustering, 2
homogeneous relations, 2, 39

IBCL, 36
individual clustering framework, 2
ITCC, 91

k-partite heterogeneous relation graph, 24
KL-divergence, 85

M-step, 109
mapping matrix, 54
METIS, 160
micro-averaged-precision, 142
mixed membership relational clustering, 43
MMRC, 43
Monte Carlo E-step, 108
Multiple-view, 47
Multiple-view clustering, 54, 127
Multiple-view learning, 49
multiple-view relational data clustering, 3
Multiple-View Spectral Embedding, 128

multiple-view unsupervised learning,
      53
multiplicative updating rules, 76
MVC, 126
MVSE, 128

NBVD algorithm, 77
NMF, 19
NMI, 150, 169
non-negative block value decomposi-
      tion, 17
Nonnegative Matrix Factorization,
      19

orthogonal matrix, 128

Relation Summary Network, 25
relational data, 1
relational data clustering, 1
row-coefficient matrix, 12, 16
RSN, 25
RSN-BD, 87
RSN-GI, 161

Semi-supervised Clustering, 118
Soft CLGA, 101
spectral embedding, 54
spectral embedding matrices, 55
SVD, 16

Unsupervised Learning with Side In-
      formation, 130

For Product Safety Concerns and Information please contact our EU
representative GPSR@taylorandfrancis.com
Taylor & Francis Verlag GmbH, Kaufingerstraße 24, 80331 München, Germany

www.ingramcontent.com/pod-product-compliance
Ingram Content Group UK Ltd.
Pitfield, Milton Keynes, MK11 3LW, UK
UKHW021612240425
457818UK00018B/513